Experiential Learning in Engineering Education

Experiential Learning presents an evolving form of education that fundamentally involves "learning by doing" and having students reflect on the work. The book discusses these recent developments pertaining to the use of experiential learning in engineering education. Covering a range of innovations in experiential learning, the book explores development in laboratories, in-class and problem-based learning, project work and society-based aspects, including Indigenous elements in the curriculum. It includes case studies and examples sourced from institutions around the world.

Features

- Focuses on recent and practical aspects of implementing experiential learning to help improve engineering education
- Offers an examination of the undergraduate experience, which leads to professional certification
- Includes a chapter on lessons in other professional education areas, such as medicine and health care, business and social work

A broad readership will find value in this book, including faculty who teach undergraduate engineering courses, engineering education researchers, industry partners that provide co-op experience and developers of training modules for practicing engineers.

Experiential Learning in Engineering Education

Alan L. Steele

CRC Press is an imprint of the
Taylor & Francis Group, an **informa** business

Cover image: © Shutterstock

First edition published 2023
by CRC Press
6000 Broken Sound Parkway NW, Suite 300, Boca Raton, FL 33487-2742

and by CRC Press
4 Park Square, Milton Park, Abingdon, Oxon, OX14 4RN

CRC Press is an imprint of Taylor & Francis Group, LLC

Library of Congress Cataloging-in-Publication Data

Names: Steele, Alan L., author.
Title: Experiential learning in engineering education / Alan L. Steele.
Description: First edition. | Boca Raton, FL : CRC Press, 2023. | Includes bibliographical references and index.
Identifiers: LCCN 2022051029 | ISBN 9780367439620 (hbk) | ISBN 9781032466910 (pbk) | ISBN 9781003007159 (ebk)
Subjects: LCSH: Engineering--Study and teaching (Higher) | Experiential learning. | Problem-based learning.
Classification: LCC T167 .S74 2023 | DDC 607.1/1--dc23/eng/20230111
LC record available at https://lccn.loc.gov/2022051029

ISBN: 978-0-367-43962-0 (hbk)
ISBN: 978-1-032-46691-0 (pbk)
ISBN: 978-1-003-00715-9 (ebk)

DOI: 10.1201/9781003007159

Typeset in CMR10 font
by KnowledgeWorks Global Ltd.

Publisher's note: This book has been prepared from camera-ready copy provided by the authors.

For Ginny, Sarah and Callum.

Contents

Preface

For a few years before writing this book I spent a considerable amount of time in an administrative role at my university. During that role I had not left teaching duties completely and I had continued to instruct on a third-year project, as well as supervise capstone projects. The administrative role included the design and overseeing of a large informal learning space housed in the library and to help develop an undergraduate research program, amongst other things. The combination of the high level perspective through the administrative role, along with the detailed focus of running a project course and supervising a capstone project gave me a broad view of experiential learning. In the administrative role I saw the impact that undergraduate research had on students. The university had showcase sessions where students could present their research to other faculty and students. I attended the US-based NCUR conference with students and saw first hand the range of work going on across North America. In the work in my home department I was already convinced about the value of experiential learning on the campus. I had designed the third-year project course and had seen how students could come up with an idea for a project and build a prototype. On returning to my department full time I decided to investigate experiential learning in engineering education further. The result of that work is this book.

The aim of this book was to collect together information to help provide an detailed overview of experiential learning in engineering that would be valuable to an instructor in an engineering program. Using theory, cases studies and examples from literature it was intended that the book be useful to new and experience faculty, as well as students and researchers in engineering education. To help with the case studies I spent some time travelling in 2019 to visit the University of Auckland and Tokyo Institute of Technology to see the developments of engineering education in countries different to my own, Canada. This was valuable and I am extremely grateful and appreciative of the time and hospitality provided that allowed me to have discussions, see facilities, attend classes and to have office space to work. Named acknowledgements are provided later in this preface.

At the beginning of 2020 the global impact of the coronavirus COVID-19 affected the day-to-day life of us all, including education. This provided educators the challenge of moving courses online and students had to adapt to working from their home or residence, with little or none of the out-of class contact that a normal campus life provided. This move to online course delivery deeply impacted experiential learning. Work had also moved online, so co-op placements became remote. Laboratories could not be accessed so the normal lab exercises students did were done at home, using remote access, equipment shipped to home, or through the use of video or simulation. Design projects became more 'pencil and paper' designs, with the use of simulation tools and limited equipment where possible. The shift to online delivery caused educators to rapidly adjust their approaches to teaching and re-evaluate the learning experience they provided. I recall writing the section on remote and videoed laboratories and being impressed with the innovation and relative novelty, compared to conventional labs. Then within weeks seeing the education system being forced to move that way for delivery. The pivot due to COVID became the central topic of many education conferences for 2020 and for the next few years, as well as education journals and departmental meetings. The impact will be long lasting.

From the outset I intended to try and keep the perspective of the book multinational, as far as possible. Hence, the visits outside of Canada for case studies. I have tried to choose examples from literature that are from around the world, but there may be still a North American focus due to the wealth of information and my own vantage point from Canada. A consequence of this is how to deal with different terminology. Here I have tried to be consistent. For example I have tried to use 'term' instead of mixing it with 'semester' and the reader should view them as synonymous. A 'course' is unit within a degree program. My own British education experience makes me aware that 'programs' are often called 'courses' in the UK and I realize this could cause some initial confusion but the context hopefully makes it clear. 'Cooperative' or 'co-op' placement is used, but again I know in the UK it can be known as 'sandwich' placement. I tried to avoid the use of terms like 'junior' or 'senior' years and instead use the year of study within a program. I also generally assume that an engineering degree program is 4 years in duration, but recognize that it could be 3 years in some countries or longer. To avoid confusion I try to use 'capstone project' to indicate the final project in a program, so a third-year project is not a final project.

I would like to give special thanks and recognize the help and support from many individuals and groups. From the University of Auckland; Mr Peter Bier, Dr Andrew Brown, Dr Bill Collis, Ms Catherine Dunphy, Dr Tūmanako Fa'aui, Dr Cody Mankelow, Mr Hugh Morris, Professor Gerard Rowe, Mr Steve Roberts and Mr Travis Scott. From Tokyo Institute of Technology; Professor Naoya Abe, Professor Jeffery Cross, Ms Yoshie Sakamoto and Professor Takashi Tomura. For the case studies in particular I would like to thank, Prof. Naoya Abe (Tokyo Tech), Mr Peter Bier (University of Auckland), Dr Andrew Brown (University of Auckland) and Prof. Robert Burke (Carleton University). For reading drafts and for discussions my colleagues at Carleton University, Mr Michael Feuerherm and Ms Cheryl Schramm. For kindly reading and providing valuable comments on the Indigenous Contributions section Mr Jason Bazylak of the University of Toronto. For visits and discussion my engineering colleagues and 3M National Teaching Fellows: Professor Greg

Evans of the University of Toronto and Professor Gordon Stubley of the University of Waterloo. For guidance on producing and editing this book Ms Kyra Lindholm of CRC Press.

Finally, I would like to thank with love my family, Ginny, Sarah and Callum, who put up with my absences whilst working on this book. It is to my family I dedicate this book.

List of Figures

List of Tables

1

Introduction

1.1 Introduction

Engineering at its most fundamental is about designing and constructing objects. It requires the engineer to understand the principles of science and then to apply them to design a solution to a problem or create something that (hopefully) helps improve the lives of people. The profession involves care in the technical details, being able to apply new technologies and ideas, communicating effectively, working within teams with diverse set of team members and learning how to develop one's own knowledge so that there can be adaption to changes in job role or technologies.

An engineering student has to learn the fundamentals of the profession within the few years of a degree program to a point to be able to work effectively in industry. The undergraduate education of any engineer establishes a base level of knowledge. Working in the profession will build on that knowledge, but establishing a solid foundation in the first engineering degree program will help the subsequent learning.

One particular area of educational development that has been growing over a number of decades, and not just in engineering programs, is the area of experiential learning. This breaks away from the transfer of knowledge, say in a lecture format, to having the student actively involved and doing something that aids the learning process. This approach to learning has been long established in engineering programs, however there has been a renewed interest in the approach, in part due to the general interest in its use across university programs. In this book we will examine experiential learning through an engineering-focused lens. The aim is to help the reader understand deeper how experiential learning works and how it can be implemented in a variety of ways in the curriculum.

In this chapter we will look at a definition of what is experiential learning, why it is increasing in popularity and orientate the reader to what is to follow in the subsequent chapters of this book.

DOI: 10.1201/9781003007159-1

1.2 What Is Experiential Learning?

At the time of writing this book, if you were to survey university websites, especially the pages of the teaching and learning centres you would find details of experiential learning and the support for this approach to learning. It has been recognized as an effective mode of teaching and is being encouraged at universities. What then is experiential learning? It is a phrase commonly used but it is sometimes not well defined. It can be varied in form and even those instructors that practice experiential learning with their students can have differing opinions on what is experiential learning. Can a classroom experience provide experiential learning, or can it only be found in projects, co-op placements or similar? In this section, we will look broadly at what is meant by experiential learning and declare our own definition from a perspective of engineering education.

The Association for Experiential Education (AEE) provides a clear outline of what they see as experiential learning [89]. They highlight that it is a teaching philosophy. Key elements are highlighted in a series of stages. There is a challenge to the students that can push them outside of their comfort zone. After the experience there is a process of guided reflection. This follow-up examination allows the students to formulate more understanding and ideas about what has been learned and what can then be applied in the future to other areas or problems. This creates a cycle as experience leads to reflection, then to ideas and to try to apply those ideas to future experiences. As we will see in the next chapter on theories relating to experiential learning this cycle used by AEE follows the ideas of Kolb and his learning cycle [150]. The AEE supports all educators that are interested in experiential education, from teachers, trainers, counsellors and therapists so the challenges and experiences their students face could be quite diverse. The role as a guide through the cycle, ensuring safety and support, as well as generally helping the learning process is universal though.

In an editorial for the *Journal of Experiential Education*, titled 'The Possibilities and Limitations of Experiential Learning Research in Higher Education' [220], Jay Roberts makes the observation that there had been a "narrow focus" in the past to what the experiential education field was and what was reported in that journal. He goes on to write about the new changes underway in the field:

> "From new curricular–co-curricular integrations, to internship programs, to active and collaborative learning in the classroom, to project and problem-based learning, there are a whole host of pedagogical approaches being developed on college and university campuses across the world." [220]

So what does experiential learning in an engineering program look like? Traditionally this may have been regarded as project work, a co-op placement and possibly work within the laboratory. However, as Roberts indicates in the quote above from his editorial there can be more, such as in-classroom activities, problem-based learning and co-curricular work. It is these broader areas, along with the more 'traditional' views of experiential learning that will be investigated in this book.

Most engineering educators will be familiar with a capstone project. Here students will have a task to undertake. It will require research, decision-making, approaches to finding a solution and creating the solution whether through hardware, software or both. It usually results in producing of an item (hardware or software) that does something and evaluation of the success should feature in the student's learning. We can see that this almost fits the previously mentioned cycle, except from an overall perspective there may be more on the experience and less on the guided reflection and abstraction. To improve the learning experience, if the model of the cycle is to be adopted then there should be stages within the project life for the student to do something and then to be encouraged to reflect on what was done and to move forward to further experiences. A project is rich in experiences, from personal research, construction, team management and communications. Each one of these can have its own smaller cycle.

What about experiences beyond the capstone project? Other commonly identified experiential learning situations are co-op placements or internships into employed positions and laboratory work. Each clearly involves an experience but there needs to be that cycle of experience, reflection, abstract thinking and further experiences. The work usually goes over a number of months and could even be a year. Within that time there will be opportunity for the student to reflect on their own work and how it is progressing. Feedback from supervisors can help with this reflection, as well as guidance on areas for development and improvement. So the experience can help the education of the student in both learning from the work, but also from contextualizing knowledge gained in courses in the classroom, as well as indicating areas where they need to learn more in future studies. Here the experience, the feedback and reflection and the subsequent work all leads through the AEE cycle mentioned above.

As mentioned, some may regard experiential learning as limited to projects, co-op, internships and similar. However, can there be experience learning in a classroom? Many traditional classes in engineering are still rooted in the lecture, which makes the student the passive receiver of information and the 'lecturer' the active transmitter. This model of teaching is long established and is still traditionally used. The difference between the experience of working on a project (as part of a course or in employment) and sitting in a lecture receiving instruction can be considerable. As Roberts mentions in the earlier quote there can be engaging challenges presented in a classroom through active and collaborative learning. Such approaches can engage a student and this will be explored in a later chapter.

Perhaps now with information more widely and easily available through the internet, lecture-based material can be disseminated in different ways so that classroom time can be used in other more active ways. This is the principle of a 'flipped classroom'. Here students gain the instruction outside the classroom, through video or readings, and then in class any difficulties in understanding can be discussed and problems can be done, all to help solidify the learning. Another approach is problem-based learning where a carefully designed problem is posed that can lead a small group of students to go and investigate the areas where their knowledge is deficient and then they all share their findings to their peers and what they learned.

1.3 Why Experiential Learning?

The benefits of experiential learning may be obvious but it perhaps it should be explored here. We know that trying to solve a problem will result in knowledge that we can use later. Whether it is solving a mathematical problem, trying to fix a piece of equipment that fails to work, or cooking a new dish, each requires drawing on our knowledge, trying to apply that knowledge to do the task and maybe having to reattempt with the knowledge of previous attempts. After a few similar situations the mathematical problem can be solved easily, the equipment or similar items can be maintained or repaired quicker, and the dish can be cooked without learning a preparation technique or knowing when to best add an ingredient. 'Practice makes perfect' as the adage says. Indeed, the educational approach of apprenticeships uses this idea of learning whilst working. Although apprenticeships can vary dependent on the trade or profession and the country in which the apprenticeship is being done, the basic principle is the same. A student who wants to learn how to do a work role learns by doing, usually guided by a supervisor or mentor and often supplemented by more formal education at a college. In engineering-related work apprenticeship education has often been used for technician education, whereas formal university-level education is required for engineers. Why was this? Part may be due to specific equipment and related skills being required by the employer. Working on a lathe or using test equipment to assess the quality control of a product may require bespoke skills for those specific tasks. Learning on the job is probably much more effective than learning from a book or lecture, being tested with a pencil and paper exercise and then finally being allowed access to the equipment. The apprenticeship approach, provides rapid and authentic training directly with the equipment, feedback is usually quick and the student/employees contribution can be incorporated quickly into the employer's work.

Why then has the apprenticeship model generally not been extended to the education of engineers? A good question as it works well for technicians and other technical staff. The range and depth of the training is probably the key factor. There is an accepted broad range of technical ability that has become traditionally the area of universities and higher education. Apprenticeships have traditionally been used for educating employees of a business and the employer wants to have the apprentice learning how to contribute to the workplace as quickly as possible. Hence the hybrid of working with experienced co-workers and the formal education. This is not to say the training is any better or worse than an apprenticeship, it is just different because of the norms and standards of the role the person is being trained for.

If we list some of the potential benefits of experiential learning we can have:

- More active involvement than passive note taking in a lecture.
- The learning can be authentic to the profession.
- Increased student engagement, including ownership of a problem.
- Often faster feedback on the quality of their work (does it work and how well?)
- Encouragement of reflection.
- Develops other skills and abilities, like communication skills or abilities to work in a team.
- Prepares for life-long learning.

The engagement of students certainly can occur through an authentic challenge. There is often more interest and enthusiasm from students with project work than in other classroom-based instruction. Attempting to solve a problem that they have been given ownership of can be a motivating factor for most. The experience of trying to solve the problem, whether successfully or not, can lead to significant learning. The student gains self knowledge and experience of how they approach their work and learning.

Often the challenge requires work in teams, briefing others, verbal explanations and report writing, all of which are important skills for an engineer. This brings us to the final point, of preparation for life-long learning. Engineering by its very nature is about development and creativity. Over a typical working lifetime an engineer will see technologies develop within their discipline as well as perhaps moving to different areas or their discipline along with promotion within an employer. All this change in the working environment can require using not only the skills learned already, but it will likely require new skills and knowledge to be developed. The individual then often has to self-manage that learning. This can be made easier from past experience so any experiential learning undertaken can contribute to that. Starting the experiential learning and its associated process of self-reflection as a student can help accelerate the learning of how the student manages their own learning.

What about the role of the educator in this process of experiential learning? At a fundamental level it can be providing the opportunity. This has to be a safe environment to learn, so there has to be the careful assessment of the safety of the learning experience. The educator can guide the student, advise and encourage the self-reflection needed to gain as much from the experience as possible. The work of Schön [231] [232] will be discussed in the next chapter but it should be noted that his work has focused on the use of the reflection and the role of the expert guiding the learner. He points out that the developed ability of an expert in a field often needs to be shown and explained to a student, who may not gain the deeper knowledge from the expert without observation and discussion with the expert.

What are the roles of institutions, like universities and employers? Part is to provide and encourage this type of learning, which is different to the traditional classroom lecture experience that has been so common for so long. The resources need to find and support the experiences can be higher than that needed for a lecture class, but indications are that there is a commitment. The more established learning experiences such as co-op and laboratories are being joined by commitments to undergraduate research, project work at the early years of a university program and innovation in the way courses are taught in a classroom. With the rise in online learning, there perhaps has been some existential reflection by institutions like universities as to what is the value they bring compared to a company or organization that provides online learning. Videos, podcasts, online discussion forums and online testing can go a long way to replicate and challenge what a traditional university has done for many years. However, the university does provide a concentration of faculty and instructors who are knowledge creators and educators. This, alongside the space, laboratories and other facilities a university has brings further value that virtual education provider will have difficulty to match. It is interesting to note and consider that at the time of writing this the 2020 COVID-19 pandemic has been ongoing for a few months and universities have been challenged by providing education with social distancing guidelines so that some universities have restricted access to the campus and facilities and learning has moved online. The advantage of the concentration of experts and students in places such as classrooms and laboratories has now been challenged with these restrictions. It will be interesting to see what the impact will be on education after the pandemic when educators have had the chance to teach courses online without easy access to spaces that were designed to help facilitate learning. My own university is about to embark on a full term of online teaching, with no use of classrooms or laboratories for teaching.

The benefits of experiential learning are many, from increased learning, the development of the individual who can become an effective professional and self-learner. Knowledge gained from other approaches can be often reinforced and placed into context better by the learner. What about any shortcomings or pitfalls? As already mentioned it does require resources, such as space, time to support the activities of all students and of course money. Perhaps the key problem is the time resource. The design and organization of the activity. Often, depending on the activity, the number of students that can undergo the experience can be limiting and so more supervisors may be needed. However, regardless of those challenges it can be rewarding experience for the learner and the instructor or supervisor to guide the students through an experiential learning activity.

1.4 What Is in This Book?

The chapters in this book are broken down in to specific learning situations, along with supporting information. The next chapter looks at some of the theories around experiential learning and looks at some of the related work on learning styles. Understanding the educational theory relating to this type of learning will aid in understanding the approaches to experiential learning and key elements of this type of teaching. This will cover a range of theoretical aspects including, taxonomies in education, learning frameworks and learning style. This will become a lengthy but necessary chapter. In Chapter 3 there will be the start of looking at a specific type of experiential learning. In engineering one of the most established is the use of laboratories and this chapter focuses on that topic. In Chapter 4 we move to the less likely form of experiential learning and that is what can be achieved in a more general classroom environment. Linking with this is problem-based learning, which is covered in Chapter 5. This is can be an overarching philosophy to education for programs and courses. Also in that chapter is a similar alternative approach to engineering education that has gained a degree of popularity is the Conceive, Design, Implement and Operate or CDIO™ approach.

Projects will be examined in Chapter 6, in both capstone and non-capstone projects, as projects become more common in stages of programs outside of the final year. In Chapter 7 the industrial experience and co-op work will be examined. This area has been another traditional area of experiential learning and has been adopted by many engineering programs. In Chapter 8 the area of experience within a university that is outside of the student's program is examined. In particular student societies and competitions, as well as undergraduate research. Many programs and disciplines have been examining the role and benefits of experiential learning in their curriculum and in Chapter 9 there is a look at three non-engineering areas, the medical profession, business schools and social work. These disciplines have similarities to engineering: they are accredited and professionally focused. Looking at some case studies on how they have utilized experiential learning may provide some value for engineering to learn from and adapt.

As experience in engineering can involve more than just technical aspect there is an examination of some non-technical areas that impact engineering. In Chapter 10 some of these areas are explored with a focus on societal aspects, including diversity in engineering, Indigenization and ethical issues. The conclusion to the book is made in the final chapter, Chapter 11, and there will be a summarizing of content and a look at some final pieces including assessment and accreditation.

2

Education Theory and Experiential Learning

2.1 Introduction

Before the different approaches to experiential learning in engineering are described it is worthwhile reviewing the educational theory relating to experiential learning. This will provide a framework for the discussions in the later chapters. It also provides details that can help with examining the theoretical aspect within specific approaches and examples that are included in the chapters.

The aim of this chapter is not to go too deeply into the theory but instead to cover some of the key ideas and concepts. If further specifics are needed they can be obtained from the references provided. The chapter will step through the core theoretical aspects which can be useful to a course designer and instructor. An engineering education researcher will be likely familiar with the details. Covering some of the fundamentals will help give context to the building of the ideas that are often used and cited in explaining the approaches to experiential learning. It is assumed that most readers will be relatively unfamiliar with education theory. In this chapter we will consider taxonomies (used in course design and development), theoretical frameworks (often used for research), models for experiential learning, learning styles that students may adopt and student approaches to learning. This should provide enough to allow consideration of how theory can help the creation and evolved design of a course, how learning through experience can be examined and how research into experiential learning can be framed. Some key areas will be highlighted as being used more frequently in this book, such as the situative learning framework, the Kolb learning cycle and the idea of reflective practice.

DOI: 10.1201/9781003007159-2

2.2 Taxonomies

Taxonomies can provide a framework for the development of a course or a program. It provides identifiable stages for a learner's progression and helps a course instructor set learning objectives. For outcomes-based learning the taxonomy becomes an essential guide to set expected learner outcomes and learner behaviours. These taxonomies provide a scaled approach to learning with tiered levels of expectations being set.

The intention is that this section does not become a detailed examination of taxonomies, but instead it is included so that those interested in developing experiential learning are at least aware of the taxonomies beyond the popular Bloom's taxonomy. A designer of experiential learning may then want to select an appropriate taxonomy and apply it throughout their design. The key factor with taxonomies is that they can provide a guide in the elements of the course, the experiences and the assessment. The choice of the taxonomy can be a fundamental decision, and as we will see there are a few different kinds.

2.2.1 Bloom's original and revised taxonomies

Perhaps the most popular taxonomy is Bloom's [36]. It has long been established and has been revised by Krathwohl and Anderson [11, 155]. In Figure 2.1, the original and the revised levels are shown for the six cognitive domains.

Cumulative Hierarchy

Bloom's Original Taxonomy	Revised Bloom's Taxonomy
Evaluation	Create
Synthesis	Evaluate
Analysis	Analyze
Application	Apply
Comprehension	Understand
Knowledge	Remember

FIGURE 2.1
Bloom's original taxonomy and the revised version by Krathwohl and Anderson.

The wording has changed in the revised version so that now verbs are used for the six levels and the final two stages of the original, *synthesis* and *evaluation*, have effectively been swapped with the revised version being, *evaluate* and *create*. These changes make for clarity. Having the culmination of the cognitive processes ending with *create* acknowledges the current view of that stage being the goal of many professional and cultural activities, whether it is creating products, business or art, for example.

Table 2.1 further explains the revised six stages of Bloom's taxonomy as defined by Krathwohl [155].

TABLE 2.1
Details of the stages of the revised Bloom's taxonomy, with associated verbs, as described by Krathwohl [155].

Taxonomy Stage	Explanation	Verbs
Remember	Retrieving relevant knowledge from long-term memory.	Recall, recognize
Understand	Determine the meaning of instructional messages, including oral, written and graphic communication.	Interpret, classify, explain
Apply	Carrying out or using a procedure in a given situation.	Implement, execute
Analyse	Breaking material into its constituent parts and detecting how the parts relate to one another and to an overall structure or purpose.	Differentiate, organize, attribute
Evaluate	Making judgements based on criteria and standards.	Check, critique
Create	Putting elements together to form a novel coherent whole or make an original product.	Generate, plan, produce

The cognitive domain of Bloom's taxonomy is often the most used and cited, but there are two other domains in this taxonomy, those being the affective domain which relates to feelings and emotions and the psychomotor domain that relates to the physical abilities or movement. In experiential learning the learning can go beyond just cognitive processes, which may be mostly used in a lecture-based class. Aspects of experiential learning may involve feelings and emotions as well as physical abilities. For example, group work can bring interpersonal challenges that can affect the feelings of members of a group and construction and assembly of a project or controlling vehicles can sometimes be needed which can require manual dexterity and physical manipulation.

In the revised scheme Krathwohl introduces the idea of a two-dimensional approach to the cognitive domain [155]. In the original taxonomy Bloom had knowledge split into *factual*, *conceptual*, *procedural* and *conceptual* knowledge. Krathwohl added another element of *metacognitive* knowledge and uses all four forms of knowledge in one axis of a matrix [155]. The other axis of the matrix is the revised Bloom's taxonomy stages. The result is shown in Figure 2.2. This matrix, Krathwohl called the taxonomy table [155], now becomes a useful tool for mapping objectives for a course, or a lesson, against the revised taxonomy. The introduction of the metacognitive category in the knowledge dimension accounts for the self-understanding of the individual's thinking processes and it is a valuable addition for experiential learning as metacognition can be achieved in self-reflection. The subject of reflection be seen later in this chapter.

The Cognitive Dimension

The Knowledge Dimension	Remember	Understand	Apply	Analyze	Evaluate	Create
Factual Knowledge						
Procedural Knowledge						
Conceptual Knowledge						
Metacognitive Knowledge						

FIGURE 2.2
Krathwohl's knowledge and cognitive dimensions matrix, known as the taxonomy table [155].

Using the taxonomy table we can now consider an example of an in-course exercise. Following a class study of a concept students move to the laboratory and undertake an experiment that uses the concept, they analyze the results and write a report about it. Table 2.2 shows how this generic engineering experiential learning exercise can be interpreted through the matrix of the taxonomy table.

We can also consider potential ways such an exercise could be developed to include elements that extend into the lower and right-hand part of the

TABLE 2.2
Example of elements of a taxonomy table for a student laboratory exercise.

Matrix element	Aspect of exercise
Understand – conceptual knowledge	Understanding what the laboratory exercise is
Understand – procedural knowledge	The experimental procedure needs to be understood
Apply – procedural knowledge	The experimental procedure needs to be undertaken correctly
Analyse – procedural knowledge	The results are collected and experimental errors are considered
Analyse – conceptual knowledge	The results are analysed to see how they compare with the concept.
Evaluate – conceptual knowledge	The overall results are presented in the report. They are discussed with reference to the concept being examined and conclusions are drawn.

taxonomy table. One way would be to have the student design and develop their own experiment, this would move the exercise into the *create* column. If then the student was to defend their experimental design and analysis choices it is possible this could extend into the *metacognition* row, although it depends on the level of the students thinking about the process.

2.2.2 SOLO taxonomy

The Structure of Observed Learning Outcomes (SOLO) taxonomy takes another structured look at learning stages and provides an alternative or complement to Bloom's taxonomy [32–34]. This taxonomy looks at the learning process and splits the competency levels in five. These are (in increasing order of competence) as follows:

- *Prestructural*
- *Unistructural*
- *Multistructural*
- *Relational*
- *Extended abstract.*

Let us look at each of these levels of knowledge structure in order. The *prestructural* level is most likely the level someone ignorant of a topic is currently at, or someone who has been assessed as failing. The *unistructural* level is achieved when a learner has a basic and simplistic level of understanding, with a single view or perspective. If the views or ideas of the topic go beyond the single level of understanding and is broader but the relationship between the elements is lacking, then a *multistructural* level is now achieved. If the relationships between the elements are understood, and there is a corresponding deeper understanding then this level of knowledge becomes *relational*. Moving to this level is also designated as entering the qualitative stages, whereas the prior levels were the quantitative levels. If the elements are identified by the learner and they are understood along with their relationships where can the next level go? This is where the entirety of the knowledge from the relational level is used to think about and abstract to help form related and connected ideas. This final level of the *extended abstract* is where reflection on the knowledge, theorizing or generalizing ideas occurs. Figure 2.3 illustrates this SOLO taxonomy and the five levels.

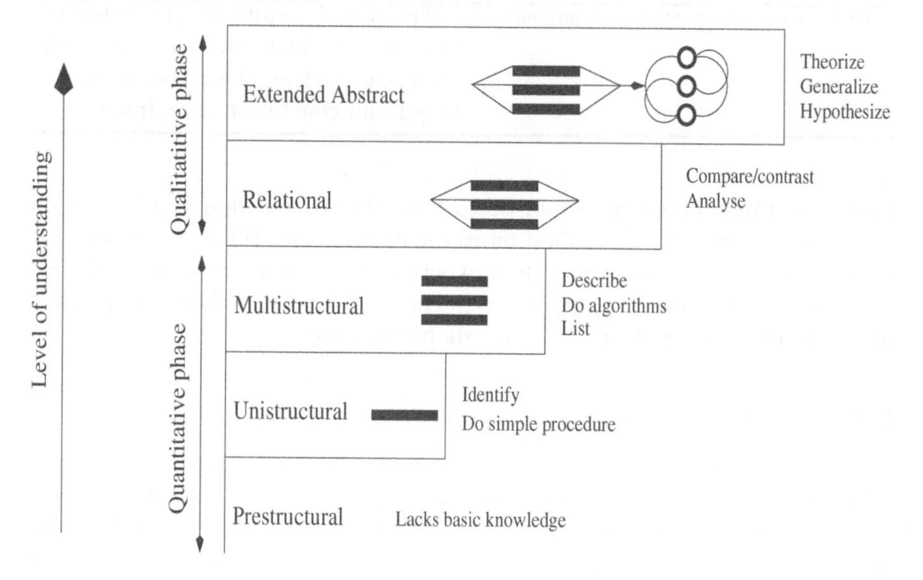

FIGURE 2.3
The SOLO taxonomy, based upon diagrams in [32] and [34].

Looking at this taxonomy and its levels, how can it help in experiential learning? Levels provide a clear breakdown of distinct stages of learning. If we return again to considering the laboratory experiment that we used previously for Bloom's taxonomy, we can see that to undertake the experiment of the class-learned concept, the student has to know at least the concept placing them onto the second of the five stages the *unistructural* stage. If the student is to successfully undertake the experiment, analyse the data and write a report on the findings, this requires a *multistructural* understanding. If a thorough understanding is shown of how the experimental method, data and final result relates together to the original concept, then we can see that the fourth level can be attained. Relating experimental methods and data analysis to a concept is perhaps different to how the taxonomy could be applied to the whole course curriculum, where at that perspective (of the whole course) the *multistructural*, elements could now be the collecting the concepts and key elements of the subject of study. In an experiment the student is often learning experimental methods, which could include the careful use of equipment and techniques, and applying them in a manner to get a result that confirms, or otherwise, a concept. So, we are applying the taxonomy to a course assessment or piece of coursework, rather than to the whole course. It does show the value of the taxonomy in that it can be applied to different levels within a program. If the student can comment accurately on the experiment and what it reveals about the concept, then they are attaining the *relational* level in that experiment and assessment. To go to the final level in the taxonomy, *extended abstract*, it would be expected that the student should be designing their own experiment and approaches to analysing the data. This may be a level that would only be expected in a final year research project or graduate thesis.

Connected with the SOLO taxonomy, Biggs also provides a relationship model for types of knowledge that is useful when working with experiential learning. Biggs highlights that knowledge comes in different forms and in Figure 2.4 we show a figure (adapted from Biggs [32]) that shows the links to four types of knowledge:

1. Declarative knowledge – textbook knowledge taught in class.
2. Procedural knowledge – knowledge of a skill or way of doing something.
3. Conditional knowledge – combining declarative and procedural knowledge together to produce something from that. For example a photography student who has studied composition (declarative) and knows the basics of how to use their camera effectively (procedural) can produce a composed and well exposed photograph (conditional).
4. Functional knowledge – extends from conditional knowledge and uses the other two forms of knowledge to provide functioning ability. In our photography example it would be being able to be a photographer.

Figure 2.4 shows the linking of these knowledge elements. Included is an example for the situation of developing computational simulations. The example shows the declarative knowledge needed to understand how to develop a simulation. Here it is likely that numerical methods, mathematics, aspects of program design and computer programming will have been covered in courses. The procedural knowledge can be the programming language syntax and the development environment. Combining those two elements allows the conditional knowledge so that computer programs can be written, tested and modified when necessary. The functional level of knowledge is now used to develop a complete numerical simulation.

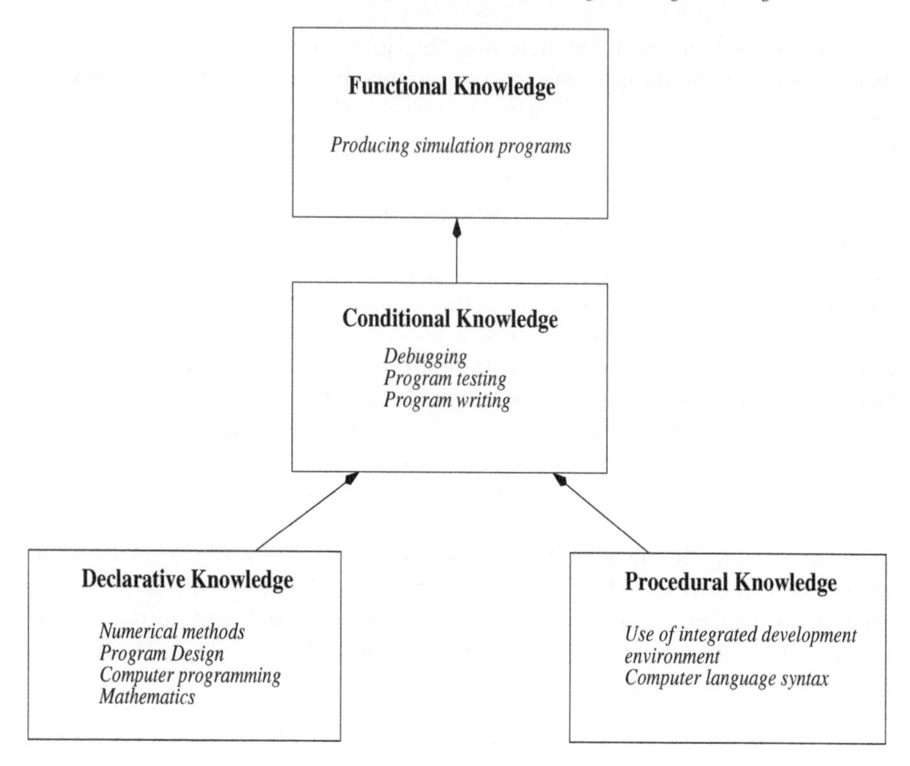

FIGURE 2.4
Knowledge elements from Bigg's relational model of knowledge [32]. Provided is an example relating to developing code for a computer simulation.

2.2.3 Fink's taxonomy

Fink's taxonomy for significant learning is the newest of the taxonomies considered here and is different to the others in that it does not consider a hierarchy to learning. Instead it looks at six elements that when combined together produces a complete and significant learning experience [97]. This approach can apply well to experiential learning, as often there is the involvement and interaction of different elements in the learning. Fink explains the approach to create this taxonomy:

> "In the process of constructing this taxonomy, I was guided by a particular perspective on learning: I defined learning in terms of change. For learning to occur, there has to be some kind of change in the learner. No change, no learning. And significant learning requires that there be some kind of lasting change that is important in terms of the learner's life." [97]

Figure 2.5 shows a diagram of Fink's taxonomy that illustrates the six elements. Straight away the inclusion of elements like 'Caring' and 'Human

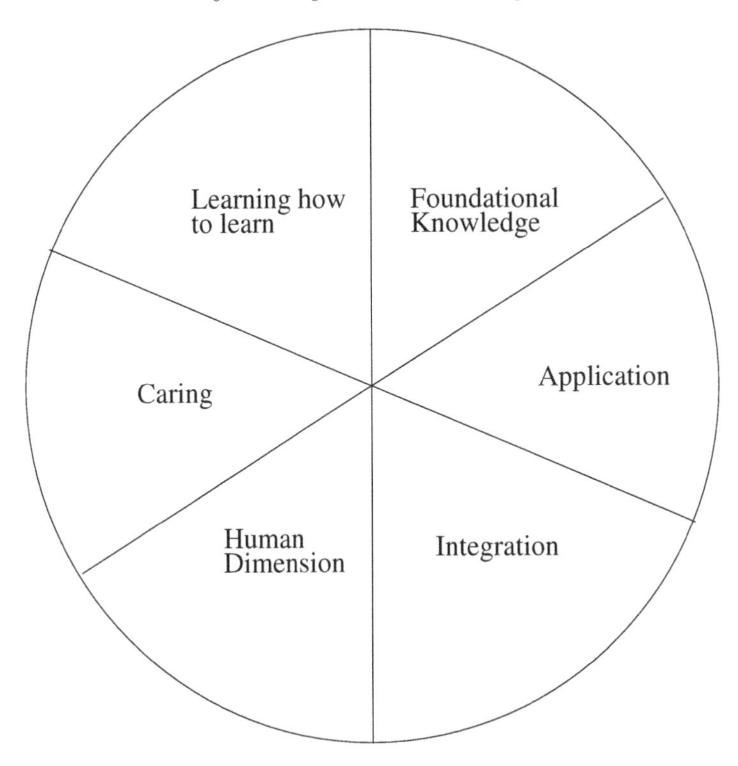

FIGURE 2.5
Fink's taxonomy [97]. Elements of the taxonomy are not in a hierarchy but can be occurring simultaneously.

Dimension' illustrates differences to Bloom's and the SOLO taxonomies. Representing the taxonomy structure as a circle indicates no priority of one part over the other elements, though for different learning experiences and courses, there could be a different emphasis on the individual elements. The use of the term 'significant learning' is important to this taxonomy and is even included in the title of Fink's book. Fink states that the taxonomy is intended to be interactive and synergistic [97].

Let us consider the elements with two experiential learning examples encountered in engineering, a laboratory experiment and a project, that are detailed in Table 2.3. The experiment could be done as part of a course and its intention is to illustrate a principle or concept that has been covered in class. The project could be a one-term group project that is a major component of a course and involves designing and constructing something and writing a report.

The lists for each example learning experience are for illustration and not meant to be exhaustive. They show how the taxonomy can be used to consider where and how learning is occurring. As can be seen from the lists in the table

TABLE 2.3
Example of applying Fink's taxonomy to laboratories and a group project.

Taxonomy Element	**Laboratory Experiment**	**Project**
Foundational Knowledge	Understanding of the key concept. Recalling previous laboratory practice.	Using ideas from the broader course. Drawing on prior knowledge from other courses.
Application	Assembling the experimental equipment correctly. Using the measurement instruments accurately.	Designing the project. Project assembly and testing. Report writing Working together and managing tasks within time.
Integration	Relating the results to the key concept. Connecting the concept in the broader context of the course.	Seeing how the project fits within the course. Understanding more of how engineering projects are done.
Human Dimension	Understanding how to use the laboratory time effectively.	Learning about working with others. Responsibility and leadership
Caring	Developing the value of making accurate measurements.	Understanding the value of working with others to achieve a larger goal.
Learning how to learn	Integrating knowledge from the class with knowledge from the laboratory. Learning to use new equipment.	Learning to share ideas. Seeing how others approach problems. Learning about leadership.

Fink's taxonomy provides a valuable way to link aspects within an experiential learning. This can allow an instructor or supervisor to develop particular areas, such as leadership learning in the project example. For a more detailed example see Dosmar and Nguyen's work detailing the use of Fink's taxonomy in developing a project following a set of junior biomedical engineering courses and replacing those courses' final exams [78]. In this implementation the authors reported "Many of the projects were creative in a way we could not have imagined" [78] as well as the project elements could be mapped to the six of Fink's taxonomy areas of significant learning, as opposed to the exam only mapping to three.

2.3 Learning Frameworks

> I take it that the fundamental unity of the newer philosophy is found in the idea that there is an intimate and necessary relation between the processes of actual experience and education. If this be true, then a positive and constructive development of its own basic idea depends upon having a correct idea of experience.
> Dewey, 1938

This quotation comes from the first chapter (titled 'Tradition vs Progressive Education') of Dewey's book Experience and Education [74]. This is a short book at a little over 100 pages; however, an important one for providing an early philosophy of experiential learning. Dewey is one of the early theorists of experiential learning, having spent decades thinking and developing ideas on education. Other famous educational theorists, besides Dewey, include Piaget and Vygotsky. Each, along with many others, influenced the ideas of how learning occurs in the 20th century and continues to influence in the 21st century.

Over time there has been a development of three fundamental frameworks of how learning can be considered. These are behaviourist, constructivist and situative frameworks [108, 200]. Each considers a different model as to how learning occurs and importantly can be used by researchers as a way to understand the learning in a particular situation. Rather than look at some of the key ideas of Dewey and others, it is perhaps most efficient for this book to look at these three frameworks. One of these frameworks, situative, we will refer to more in this book than the others, as this framework may be more valuable to use to understand experiential learning.

2.3.1 Behaviourist

Behaviourist theory works on the principle of observation and feeback on the seen behaviour in a given situation. The stimulus of the situation should elicit a response. There may be a correct response for the particular situation and desired behaviour. Teaching and learning can focus on showing and educating about the required response and then the assessment can be made by checking that the correct response is given. For example, a student is shown how to solve equations numerically. The student then has to produce computer code that solves a particular equation. The assessment of the student's code becomes important for feedback. If the checked code is poor or incorrect there needs to be feedback such that there is a change and an improvement in the future.

For example, a low grade is assigned if the response to the exercise is poor or incorrect, or a higher grade is given depending on the degree of success. The grade then becomes the stimulus to improve. Correct knowledge or behaviour is reinforced and hopefully strengthens the learning. An incentive is provided if the knowledge is inadequate, with future learning focused on improving the behaviour

The process is based on stimulus and response. Feedback comes from the type of response. If we continue with the student learner example, the feedback will continue until the appropriate level of desired ability is shown and the student passes on to work with the knowledge acquired and can progress to the next level of learning. This process of observation, feedback and adjustment is the basis of the behaviourist model. It is motivated through an external process and does not consider what is happening internally in the individual's mind. It is the individual path taken through the observation, progress and feedback that each learner follows which makes their progress unique.

2.3.2 Constructivist

If the behaviourist model is externally motivated then a cognitivist approach is internally motivated. This framework uses the idea that each individual develops their own model of how to understand something. Learning then becomes the comparing of that understanding to the new learning situation and the adaption of the models to make comprehension of the new situation. Memory of the models and the situation is required as well as the processing of the models to understand the new. If there is limited exposure to previous models, that is a lack of experience, or misunderstandings or misconceptions then the learning can be hindered. Successful understanding of experiences can provide the building blocks for future learning. This approach requires the learner to actively engage with the material to be learned. Passivity and not engaging with the experience will inhibit understanding.

Construction of accurate models of knowledge is key. Inaccurate models require the model to change and a new and accurate one to be developed. New learning is building upon the pre-existing models of understanding. The active engagement with the work can be done individually, or there can be interaction with others to consider and compare their understanding and models of how a situation is working. In experiential learning this model development and application could be the use of prior laboratory experience to understand how to do the laboratory work in a new course. Or, it may be the use of the experience of effective communication skills in a co-op position being used in a capstone project to help in report writing or presentations.

This area can have sub-theories [200], such as *information-processing theory* where an analogy to learning is made to the operation of a computer, with access to memory and processing; *social cognitive theory*, where learning is done by observing others; and *constructivism* where an individual builds up their collection of models. Their collection and understanding are unique to them. Metacognitivism can even arise from an individual's understanding of their learning approaches and goals.

2.3.3 Situative

If behaviourism is perhaps the oldest of the three frameworks, situative is the youngest. The origins of this framework are usually attributed to the work of Lave and Wenger [158]. The essential basis of this approach to understanding the development of learning is that knowledge is distributed across a community of practitioners and a newcomer to that community often starts at the periphery of the community and through learning moves deeper into it. Members of the community pass on knowledge. Tools and specific language are used by the community. Through participation in the community students can start to acquire the identity of a practitioner, as knowledge, specialist language and ability with tools are learned. As can be seen from this brief outline, there is strong applicability to experiential learning as well as to the education of an engineer [141, 200].

Situated learning can be broken down into three key elements: *Action and Interaction*, *Mediation* and *Participation and Identity* [140]. With action and interaction, here the learning occurs through doing relevant activities and working with and under the guidance of those with the knowledge. For example, being shown how to work with equipment on a test bench and then making the measurements successfully on their own. This example can lead to the second listed element, mediation. Here the learner needs to know how to use the equipment on the test bench correctly. They will need to record and present the data correctly and even be required to interpret the displayed data. This is all mediation, whether with equipment, diagrams, graphs, computer languages, mathematics and other semiotic forms.

The third element is participation and identity, which comes from doing the activities and using equipment, diagrams and analysis techniques, which develops and deepens the individual's identity of (in this case) being an engineer. The accumulated learning of the student moves them from the periphery of the community to a deeper level within it. This can strengthen the interactions and exposure to other activities and tools, further building identity. If we consider how undergraduate students interact in laboratories on campus, there can usually be seen to be a considerable development over the years of study, such that the interaction in the capstone project can be quite different from their first experimental work in their first year. The identity builds within the learner and the interaction with the others in the community can change. An engineering student may also join engineering societies and participate in other campus activities, such as engineering design competitions or undergraduate research, which can further increase the identity of becoming an engineer. The community may have 'rituals' such as graduation with an engineering bachelor's degree, to becoming an engineer in training, to becoming a licensed professional engineer.

Exposing students to practitioners and expecting students to learn and take on the identity of learners is not as simple as may be first thought. Sadler gives an argument in science education that a student who is trained to undertake some complex scientific task with equipment does not automatically become a scientist [228]. Brown, Collins and Duguid also stated [46]

> Unfortunately, students are too often asked to use the tools of a discipline without being able to adopt its culture. To learn to use tools as practitioners use them, a student, like an apprentice, must enter that community and its culture. ...learning is, we believe, a process of enculturation. [46]

This raises an interesting question as to what is the culture in a learning community? Whether it is a broad community like a profession, or smaller like the environment in a laboratory or classroom. The anecdotes an instructor may recount to students of past engineering work can add to the culture. So the culture, interactions and participation can be more complex than first thought. There can be 'peripheral participation' where a student may be learning by observation of a situation that they are not directly participating [46]. For example, observing what other groups are doing with their experiment in a laboratory room.

Unfortunately, communities may not be ideal for effective learning and practice as Macpherson and Clark investigated [165]. Their study showed that different work groups of a utility company in the UK could have different rates of causing damage to other utility services when digging. The knowledge of the job in the work gangs was passed on within the gangs and it was seen that some gangs with members with the longest service did not always have the lowest damage rate on other utilities. In short, poor practice was being passed on and mediation through systemic organisation prevented the cross flow of knowledge to reduce the strike rate. This produced what Macpherson and Clark termed 'Islands of Practice' [165].

Situated learning provides a way to consider learning and its efficacy from a perspective of a community. In an undergraduate learning laboratory, there will be a room with equipment, diagrams of the experimental setup, measurements that have to be made, student partners, other student groups, teaching assistants, support technicians, instructors and experiments to conduct and report on. This example shows the situation-rich environment that learning occurs within. This framework is significantly different to the cognitive framework previously mentioned, where everything is considered internal to the mind of the student, who has interacted with the learning situation (for example, undertaken the laboratory experiment). A blended framework of cognitive and situative approach can be considered [200], where situations with with authentic learning activities can help develop the cognitive and conceptual models a student has of a particular system. Again, we can return to the laboratory experiment example, where a particular concept is being tested in the experiment and the learning is occurring in the laboratory room as previously described.

Because situated learning provides a rich and useful way to consider experiential learning it is for this reason it will be the preferred framework used within this book. That is not to say that other frameworks are not valid to use experiential learning and may be used in the book too.

Having considered different learning frameworks, we will move on to look at some specific theories related to experiential learning, Kolb's learning cycle and Schön's reflective practice.

2.4 Kolb Learning Cycle

Perhaps the most well-known theoretical work relating to experiential learning is David A. Kolb's 1984 book *Experiential Learning: Experience as the Source of Learning and Development* [150]. Described in this book is a cyclic four-stage model for experiential learning. This cycle contains the stages of concrete experience, reflective observation, abstract conceptualization and active experimentation, Figure 2.6. This model has become popular for considering the learning process occurring within experiential learning.

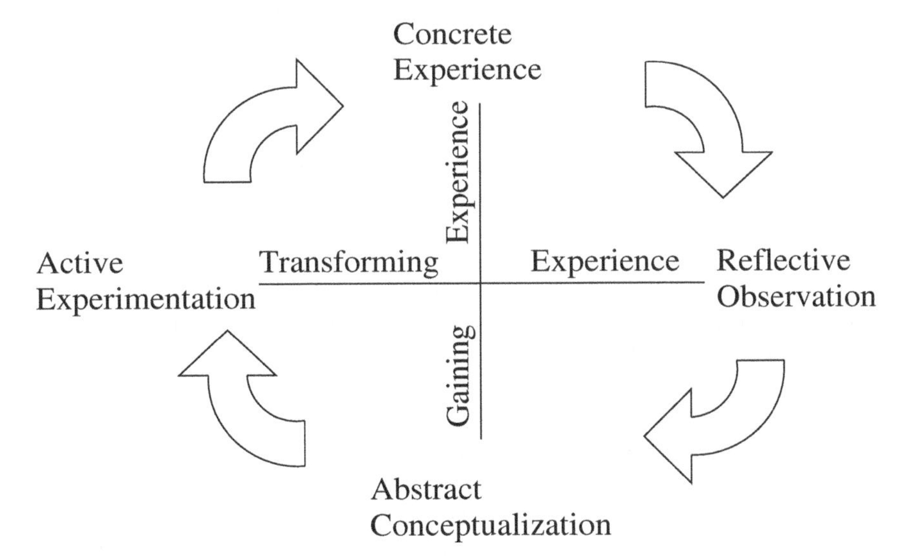

FIGURE 2.6
The Kolb learning cycle for experiential learning. Detailed in [150]

The model's four stages make up the learning process, which can start at any one of the four points. In experiential learning perhaps the most frequent starting point is the concrete experience. For example an experiment is conducted by a student to heat a metal bar, and temperature is measured across the bar from the heated end to the unheated side. Readings across the bar are taken and the data obtained can be examined. Perhaps there is a pattern and trend seen. The student in making these observations has now moved to the reflective observation stage. Having collected these observations how does the data match with what is known about heat conduction in metals? This can move the learner to the abstract conceptualization stage. Knowledge from classes and books can be considered and now we are at the abstract conceptualization stage, as we try and understand what is occurring from a physical model of heat conduction in a metal. From the combining of a theoretical understanding with the collected data further experimentation can be considered. What if the bar was cooled? What if the bar was a different metal or a different material like a plastic? Now we are in the active experimentation stage. This could lead to another experiment, which becomes another concrete experience and the cycle can now be repeated.

Having described the cycle it is worth looking a little deeper at how information is processed. Kolb discusses the two ways of gathering knowledge, through concrete experience and abstract conceptualization and also the two ways of processing the knowledge, through either reflective observation or active experimentation. First, we can look at the two ways knowledge is collected, the concrete experience and abstract conceptualization. This is the 'gaining experience' axis shown in Figure 2.6. Kolb points out in his book "Two distinct modes of grasping experience may not be readily apparent, but it is a fact that can be easily demonstrated with but a little effort" [150]. The example demonstration Kolb gives is simple but effective. He asks the reader to consider their surroundings. The chair they are sitting on, the room they are in and what the reader sees and feels. This is concrete experience. He then poses the situation that if the reader was to leave the room then although there is no direct sensory experience of the room once it has been left, the reader could still describe the room using comprehension of it and its contents. The ability and the detail of the description depend upon the reader's communication ability. Now if we consider a similar demonstration of these two ways of acquiring knowledge in terms of engineering. A student could assemble a simple electrical circuit or a wooden structure. The outcome of the circuit or structure can be directly observed. A light could come on when a switch is closed, or a weight could be applied to the structure at different positions until the structure starts to unbalance. If the student then had to describe the circuit or the structure and its corresponding behaviour, then there would be the use of words, numbers, mathematics, diagrams and graphs. Of course an engineer's education requires the development of the various ways to communicate accurately through the development of writing skills, mathematical analysis, circuit diagrams, technical drawing, engineering models and graphical analysis. Indeed we can see now how undergraduate education of engineers involves a flow between the two stages of concrete experience and

TABLE 2.4
Kolb learning transitions. Detailed in [150].

Label	Transition
AΔI	Apprehension by intention
AΔE	Apprehension by extension
CΔI	Comprehension by intention
CΔE	Comprehension by extension

abstract conceptualization. Students in classes can learn the concepts and the associated communication of ideas (through mathematics, models, diagrams, etc.) and then can experience directly in the laboratory a specific case through construction and measurement. The acquiring knowledge through concrete experience is called *apprehension*, whereas knowledge gained through abstract conceptualization is called *comprehension*.

So, what about the process of transforming experience (the horizontal axis in Figure 2.6)? This occurs with the stages of reflective observation and active experimentation. In the observation stage there has to be consideration of the experience. It could be looking at collected data and perhaps plotting it, to continue with the experiment example. Whereas, the active experimentation stage is the decision process of what to do next. Now there is consideration of what to do next. Processing of insights from the earlier stages in the cycle can lead to a choice of another experiment or course of action. Like the gaining of knowledge through *apprehension* and *comprehension*, the processing or transformation of knowledge has labels too; for transformation through reflective observation the processing is done by *intention* and for active experimentation the processing is done by *extension*.

The Kolb experiential learning cycle can at first sight be deceptively simple and straight forward in operation; start at one of the four points on the cycle and proceed clockwise around. However, as Kolb shows in the book the individual learner can move in different ways around the cycle and will likely have favoured paths or quadrants of the cycle. We will talk about learning styles a little later, but here we will discuss the process first. The complexity of learning for individuals can then be seen through the model. Moving from the two prehension stages, *apprehension* if knowledge comes from concrete experience, or *comprehension* if the knowledge comes from abstract conceptualization, can be done through *intention* (via reflective observation), or through *extension* (with active experimentation). These transformations or learning strategies are labelled as shown in Table 2.4.

So if we look at say AΔI then this indicates the learning strategy of apprehension by intention using concrete experience and reflective observation. Similarly, CΔE looks at the learning by comprehension by extension through abstract conceptualization and active experimentation. It can be seen that each of the quadrants of the Kolb learning cycle has a learning strategy linked to it.

Having learning through all four of the processes will provide a deep learning, but it is pointed out by Kolb, with examples, that individuals will favour certain learning processes and consequently have preferred learning styles, which will be discussed later in Section 2.6. As well as an individual's preferred learning style Kolb also highlights that an individual's professional education and developed habits can also favour certain learning approaches. Kolb notes from surveys that engineers generally favour abstract conceptualization and active experimentation, so acquiring knowledge through *comprehension* and transforming it through *extension*. Whereas a social worker may generally favour the style of concrete experience and reflective observation [150], pp. 88–89.

Applying this to the formal education of engineers it can be understood that the classes and resources available can enhance the comprehension of the student engineer by the learning of engineering concepts, physical models and mathematics, or learning by *comprehension*. Whilst laboratories, projects and industrial experience providing rich opportunities for concrete experience, learning by *apprehension*. Encouraging the transformation through *extension* or *intention* allows the development of knowledge. Of course for an engineer, like all professionals, the learning does not stop after graduation and learning will continue as the engineer practices their profession. Experience can strengthen the learning ability and the process of moving around the cycle may be faster. When faced with findings from an engineering project the process of reflection, linking to concepts and onto decisions for further work or experimentation may be faster for an experienced engineer than for a new graduate. This ability to reflect and action from that leads us to the next learning model, Schön's ideas of reflective practice.

2.5 Schön's Reflective Practice

In 1983 Donald Schön, a Professor of Urban Studies and Education at MIT, published his book "The Reflective Practitioner" [232]. This brought together his ideas of the need and encouragement of reflective practice in professional work. He criticized the focus professional education had at that time on teaching systematic knowledge so that when a well-defined problem occurs it can be solved by application of that knowledge. Schön uses the term *technical rationality* to apply to this systematic use of knowledge. He argued that if the problem is poorly defined, ambiguous or has even conflicting factors (such as technical, economic or ethical) then a practitioner could have a challenge to solve these problems. However, accomplished practitioners, experts, are often able at tackling and solving such challenging problems, in a way that could be described as artistry. How then can this 'artistry', drawn from a wealth of experience and reflection, be used and importantly be passed on to students and others so that other poorly defined problems can be solved? In moving from the rigorous application of knowledge to solve the well defined problems he proposed a move from just *knowledge in action* to the incorporation of *reflection in action*, to develop the awareness needed to solve these so-called vague and poorly defined problems. This then becomes the route into artistry in the profession. In proposing this it raises the point that mastery of a subject often develops in an individual, so that potentially the knowledge the individual has becomes challenging to communicate. For example, how can a thorough technical knowledge of music be developed into the mastery that a skilled musician can possess when playing, and then how can that mastery be communicated? Similar mastery can be exhibited in engineering for example in resolving technical problems encountered in a simulation, an experiment or on a construction site. The resolving of the problems can rely on a combination of experience, a deep technical insight and for some, it is a developed mastery that others will want to learn from to help develop their own knowledge. In engineering this degree of mastery often leads to promotion and ultimately into managerial-type roles. At a senior level the master engineer will have more engineers reporting to them and so can potentially help lead that team to solve more problems, or effectively guide through the avoidance of problems. Of course, the senior engineer is more abstracted from the problem and so may not be asked to advise or intervene until the problem has escalated to a serious degree. Therefore, if there was an effective way that the mastery or even 'artistry' of the managing engineer can be communicated to the other reporting engineers, then the mastery level of the group is multiplied. Transferal of experience and the many small aspects that knit together to give a masterful practitioner their ability is not easy or straight forward. Schön indicates that mastery that is drawn upon is used in a holistic way, there are technical knowledge, past experience, reflection on that past experience and even the building of their own theories. How can this holistic 'education' be passed on to another to transfer the mastery? The development of reflection-in-action and the use of coaching is discussed in [232] and further in [231] and Schön provides detailed examples in architectural design, music and psychoanalysis. Others have applied to the ideas to engineering [80] and engineering education [266].

Development of the reflection in practice can be done through encouraging the individual to reflect on their own and to use a coach to help with the reflection. Presumably in many situations the coach for a student is a faculty member, but it could be a teaching assistant or a fellow student, for example a fellow group member. The advantage of using a coach is the student can be directly guided with the reflection and that guidance can help the student move forward with problem drawing on the experience and mastery of the coach. This however can bring problems through the way the coach and the student interact and Schön goes to some length in his book *Educating the Reflective Practitioner* to show the interactions [231]. The interaction between the student and coach becomes a three-way interaction. The student and the coach both interact with the problem and then the student and the coach interact together. The problem, student and coach are three points in the interaction. So the two individuals may have different observations and ideas on the problem and its solution and then these two people can share their thoughts on the problem together. This three-way interaction needs to be carefully done to help in the student's learning. If either the student or the coach misunderstand each other then there can be an impediment to the reflection and ultimately to the learning. As Schön points out, the responsibility of the communication and interactions is on the coach and how the coach interacts with the learner can become important for the interactions related to not just the current problem but future problems. Drawing on his earlier work with Argyris [13] Schön discusses two types of interaction [231]. One approach that could occur is for the coach to take the opinion that their ideas are superior which is clearly communicated in the interaction with the student. Then one or both in the interaction could take a defensive position in the reaction to their ideas on the problem. Control of the problem is attempted by the coach and this can limit the choice in resolving the problem under discussion. Schön labels this a type I interaction. In the second or type II form of interaction the coach does not become forceful or defensive of their own idea but becomes more of an equal to the learner so the problem is more jointly owned and shared. This can encourage the student to explore and discuss the problem, guiding the student towards their understanding of the solution. In terms of the length of the relationship of interactions there can be a tendency for type I interactions to be brief and last little beyond one problem interaction, whereas the type II interaction can last longer and will more likely be repeated because of the shared interaction. In practice we may recall interactions where a solution to a problem is directly proposed by the instructor in a manner that stifles exploration and results in defensive positions. This can reinforce that there is only a single correct solution that inhibits the learner to reflect on the problem, despite the possibility that there could be a range of possible solutions. It is also reasonable to expect that once the interaction has occurred the student will likely not be keen to return to the coach for future assistance. Of course depending on the circumstances of the problem, if there is a safety issue or any health risk to the student or others the coach may need to show their concern and provide a direct strategy for safety.

Schön criticized the approach of technical rationality in the training of professionals. The concept of technical rationality being the rigorous application of learned professional knowledge to solve a particular problem. The education system Schön argues for many professions is routed in developing the technical knowledge, often in a scaffolded way, such as medical professionals being first trained in the basic sciences before moving onto content that provides more applied knowledge of the profession. He argues that this leads to challenges in solving some problems that include vagueness and uncertainty in them. It is in this way he argues the case for professionals to undertake reflective practice and to draw on tacit professional knowledge when working in a professional capacity. The question that can then be raised is how does an individual acquire this tacit knowledge and learn how to enhance its application? This leads us back to the Kolb learning cycle and the process of learning from experience. Kolb has reflection directly following concrete experience in the learning cycle. There has to careful consideration of the experience before attempting to proceed further, especially if the problem is challenging. If a practitioner keeps reflecting on their work then successful solutions could pass into becoming tacit knowledge, such that the path to solving future similar solutions need little consideration - the artistry that Schön described. There is a connection between the ideas and models of Kolb and Schön through the reflection. Both deem this to be an important part of experiential learning. So when students are exposed to experiential learning and their reflection on the experience is nurtured and encouraged, then students are put on the path of becoming reflective practitioners.

2.6 Learning Styles

So far we have looked at models and theories for how to design learning experiences (taxonomies); how to interpret how learning is occurring (learning frameworks) and processes to understanding learning, especially experiential learning (Kolb and reflective practice). What about how individual student's receive and respond to learning activities and approaches? For this we will now move onto looking at work done on learning styles and approaches to learning.

2.6.1 Learning styles and Kolb's cycle

Previously in Section 2.4 Kolb's learning style was covered. As mentioned then this can lead to the idea that individuals will have a preference to learning in one of the four quadrants. The quadrants being defined by the way information is obtained and then by how it is processed. These styles have been given names relating to the quadrant that the individual prefers to use when learning. The names in the related quadrants are shown in Table 2.5 and these are *diverger*, *assimilator*, *converger* and *accommodator*. Divergers use concrete experience and process with reflective observation. Kolb indicates divergers are imaginative and have a strength towards viewing situations from different perspectives and seeking meaning from the various perspectives. Assimilators use reflective observation to process abstract concepts. Convergers use active experimentation to process abstract concepts and accommodators use active experimentation to process concrete experience. None should be viewed as a better form of learning compared to the others here, but it can be considered that students of particular programs or disciplines may have a higher number in their population that favour a particular learning style. The instructor in a course may also teach in a manner to favour a particular learning style. For engineers, studies have shown that there can be a predominance of convergers [150], p. 127 and accommodators [150], p. 167.

TABLE 2.5
Kolb learning styles, with some traits and strengths as well as likely professions. Derived from [150].

Learning Style Type	Traits and Strengths	Likely Profession
Diverger	Imaginative and can view situations from different perspectives and can seek meaning from them.	Artist
Assimilator	Inductive reasoning from observation and use of theoretical models.	Scientist
Converger	Problem solving and making decisions. Solves problems by applying ideas.	Engineer
Accommodator	Seeks opportunity and adapts. Takes action and carries out plans.	Manager

To try to determine what an individual's learning style is in the model of Kolb's experiential learning model Kolb developed the Learning Styles Inventory (LSI), which after version 4 became the Kolb Experiential Learning Profile [90]. In this version the original four learning styles are changed into nine styles. Each previous four quadrants are now divided into eight and a style is created for learners that show no signficant preference to either of the other styles. So from the quadrant structure this style, called *balancing*, would be centrally located at the intersection of the quadrant lines. The styles are [91]:

- *Initiating*
- *Experiencing*
- *Imagining*
- *Acting*
- *Balancing*
- *Reflecting*
- *Deciding*
- *Thinking*
- *Analyzing*

Over the years the LSI has been developed further and as mentioned the current version is now called the Kolb Experiential Learning Profile [90]. The inventory is ultimately a tool to determine an individual's learning style in a specific model for learning, in this case the Kolb experiential learning model. That style may change over time as further learning and personal and career development occurs.

For an instructor of a course the question can be raised should the learning styles of their students be known. This will require an investment of time and possibly money (as some inventories have to be paid for, like Kolb's LSI 4.0). Perhaps the most important points an instructor needs to consider are:

1. Students can have different learning styles. If you want to use Kolb the four basic descriptors are *diverger*, *assimilator*, *converger* and *accommodator.*

2. Due to the prior education and their decision to select a certain program of study there may be an uneven distribution of learning styles. For example, in engineering there may be more students with *converger* learning styles than *divergers* or *assimilators*, if the Kolb model is being used.

3. Regardless of the distribution of learning style, it maybe more effective to ensure there are a range of learning opportunities that creates different learning experiences. Again using the Kolb learning cycle the process of moving around the loop during a course can be important for the development of all the student's learning, regardless of style.

4. At the individual level, any student that struggles with a course, it could be an indication that the course is providing few learning opportunities in a manner that supports the individual's learning style. Again another reason for point 3.

5. During an individual's progression through a learning process, such as a degree program, there could be more of a move towards a particularly reinforced learning style. This can continue as life-long learning continues. It also raises a question is the learning style that is being reinforced by the profession now the style that will be needed in future for the individual (e.g. an engineer that becomes a manager) or for a profession?

2.6.2 Felder and Silverman's model of learning styles

Besides the Kolb experiential learning model that leads to a categorization of learning styles, there is another popular model for learning styles which was proposed in 1988 by Felder and Silverman [94]. Here four different dimensions are considered and a learner can be assessed as to where on a scale the learner can be placed, with two labels being applied to the two poles of the spectrum of that dimensions. This is somewhat similar to what the Myers–Briggs Type Indicator uses. The dimensions are perception; input; organization; processing

TABLE 2.6
Felder and Silverman's categorization of learning styles. From details in [94].

Dimension	Poles	Explanation
Perception	Sensory – Intuitive	How a learner prefers to perceive the world. From sensing the world through data and experimentation or using models of theories to form an intuitive view of the world.
Input	Visual – Verbal	The way the learner prefers to receive information, visually through diagrams, pictures and symbols or through words and sounds.
Processing	Active – Reflective	What a learner does with the information. From actively using the data to discuss or conduct tests. Through to reflectively examining the information and manipulating it through consideration.
Understanding	Sequential – Global	Whether the learner can learn a subject step by step as there is progression through the information or whether the information can only effectively be processed when the majority has obtained and can be looked at as a complete set.

and understanding. Table 2.6 shows the names for the opposite halves of the dimensions

Before proceeding it is worth noting the original paper [94] included five dimensions, the other being the organization of information, either inductively or deductively. Here, organizing information and moving from principles to applications is deductive and commonly used in traditional classes. Alternatively, inductive moves from phenomena and observations to inferred principles. Felder notes that inductive learning is a better approach for students, though challenging. He then prefers to drop the organization dimension from the model. His preface written in June 2002 on his paper, located on his website, gives details and his argument [94].

As experiential learning fits with his view on the use of inductive learning then it seems appropriate to adopt Felder's subsequent recommendation to avoid the use of the organizing dimension. Using four dimensions there are then 16 possible combinations of the poles that can be used to help describe the learning style of a student. As well as the student's learning style the paper includes the teaching style of the instructor which is formed in a similar way. Assuming that a teacher will intentionally be changing their style of approach and that in an experiential situation a student can have learning opportunities coming from different sources than just an instructor, the focus here will be on the learning style of the student. In Table 2.6 there is a basic explanation of the polar ends of the dimensions and learners will have their individual learning style located somewhere between the two poles of each dimension. The model again clearly shows how individuals can differ in their approach to learning. In experiential learning these differences in learning styles can become more apparent as students respond to less well defined problems, often with real world information and data and have to interact with peers who may have a different learning style to each other. This richness in the learning opportunity can however help an individual learn more about the process of learning and their own approach to problem solving, compared to others.

Let us consider now the poles of the learning dimensions in relation to experiential learning. With the perception dimension students tending towards sensing may appreciate the access to data and facts arising out of the learning situations. Intuitors may prefer to consider the diagrams, models and theories with the situation being analysed as well as the creativity of the problem. In short, students at either end of the poles of this dimension will approach an experiential learning situation from either a concrete analysis or abstract examination of the problem. An instructor may have a preferred route if they are guiding students through the problem, however, there can be valid approaches from either pole of the dimension.

With the input dimension there can be implications on how students may want to gather and examine data on the experiential learning activity. Those with a preference towards visual learning will want to have information in or transformed into visual form, such as photographs, graphs and diagrams. Whereas those with an auditory preference will perhaps prefer to have material to read or they may want to discuss the activity and to what is to be done.

The processing dimension has the poles of active and reflective. Here the two extremes could lead to some differences of opinion to how to approach an experiential learning exercise. For example, if two students with the two different styles of this dimension were given the task of learning a new complex piece of equipment, the active individual may want to try to get the machine working and learning from that, whereas the reflector may want to read the manual and plan some specific tests when comfortable with understanding the principles of the machine. This situation can sometimes be encountered in a laboratory setting with student partners working together on a laboratory exercise.

The understanding dimension provides an interesting set of poles and corresponding approaches. A sequential learner may want to plan a progression through the experiential learning exercise, mapping stages and sequenced targets, whereas a global learner will want to jump to specific stages or solutions that they consider important. This diversity of approaches (like active and reflective learners) could lead to conflict within a group if a single approach has to be selected.

2.7 Approaches to Learning

Having now considered a student's possible learning style(s) we should also consider their possible learning approaches. These approaches are different to learning styles as will be seen the approaches a student can take can change depending on factors such as the course or the method of presentation of material. In 1976 Marton and Säljo published work that looked at the different 'levels of processing' of information that was provided to students [171]. The study assigned passages of material to students, who were paid volunteers, and they were asked to study with a recommended limit on the study time. They then were tested on the passage as well as more open questions about how they approached the reading of the passage. Six weeks later there was another test to look at retention. The changes in the response could be compared. The author's reported finding two distinct approaches to learning the material, the first was a reading and understanding that could reproduce the text, more of a rote learning, which they called *surface-level* processing. The other approach was that some students tried to understand the intended meaning of the author from the text and this they called a *deep-level* processing. Another experiment was also undertaken by the same authors and revealed similar evidence of the two types of learning, as two groups of students were either factual-based questions or meaning-focused questions [172]. Results showed evidence of an adaption to the type of learning once there was a familiarization with the type of questions asked.

The term 'processing' was mentioned previously and this was discussed as a proposed amendment to the term 'approach' made in later work by Entwistle, Hanley and Hounsell [86]. In their paper they report on the development of an inventory of approaches to studying, when considering that learning has the aspects of *intention*, *process* and *outcome*. The change of wording was to avoid confusion with the word 'process' that was also used within the three stages. Besides a clarification to the naming, an important contribution of the paper was the addition of another learning approach, that of *strategic*. This is a form of learning that focuses on identifying what the examiner or assessor's approach will be in determining the student's knowledge. Entwistle et al use the terms 'cue seeking' and 'playing the system' when describing the approach students may take to determine the rules of assessment and use them as a path for their own learning [86]. Now we have three identified approaches to learning, *deep learning*, *surface learning* and *strategic learning*. Of course it is the first of these three, deep learning, that educators are wanting to encourage learners to adopt.

We should now consider these three approaches to learning when applied to an experiential learning. Having personally run a group project class for a number of years there have been some certain traits that can be seen in students adopting different strategies. It is important to note that students may adopt different learning strategies at different times. If the material is of interest to a student then a deep learning approach may be taken. Alternatively an uninterested student with a busy schedule may take a strategic approach for just one of their courses. In short these are not a generalization of a student's ability, at least not solely, but more of an indication of a student's interest, motivation and even health.

- A deep learner is most likely to be seen as spending time on the task. Promptly attending seen to diligently working, often collecting and reporting information. Their questions will often reveal the intent to gain a fuller knowledge of what they are working on. In group presentations or report writing they may take a lead role as they may feel they have a fuller understanding than others.
- Surface learners may show a lack of punctuality and more frequent absences. Progress on tasks may be slower than expected. In group work they may let others do the major part of the work. Attention on the tasks may be seen to be lacking, perhaps spending more time talking with others. Questions from the student may focus on the route to finishing a task. There may be a general lack of questions too. If group work is involved others may become irritated by the unreliable nature or lack of productivity of the surface learner. In group presentations they may defer to others to lead.
- A strategic learner may show similar tendencies to a surface learner. Absences from sessions that do not directly lead to graded work. Sometimes there will be a focus on work on a deliverable. Impatience or a focus on getting a decision made quickly rather than by careful deliberation may also be a sign. Questions may focus on what is required for assessment and clarification on how grading is done.

Now, these are generalizations and there could be other factors at play too, such as an individual's tendency to introversion or extroversion or even health issues. An instructor who is often overseeing the learning activity can soon start to see trends though and some direction by the instructor may be needed. If there is no oversight and the student works on a task away from the observation of the instructor there can then be a difficulty in determining the learning strategies. It should be pointed out too that if group work is involved the different learning strategies taken can be a source of tension within the group. Absences, lack of productivity and quality of work can all result in disharmony within the group. The surface and strategic approaches of some students can be one reason why group work is unpopular with students. An instructor may want to consider some type of expectation guidelines for group work. Later in Table 6.2 of section 6.5 I provide a set of questions I have developed and used for group work that has evolved to help avoid issues within groups.

2.8 Discussion on Reflection

We have examined ideas and theories on student learning as they encounter new experiences and problems within those experience. A route mentioned by more than one theory is the use of reflection. On the surface that may seem a straight forward task to undertake but perhaps we need to look a little deeper into what a reflection is and then look at how this has and can be used in engineering.

First let us consider a basic definition of a reflection and here we start to encounter some of the challenges associated with the term. In 2002 Rodgers states in the paper 'Defining Reflection: Another Look at Dewey and Reflective Thinking', "In fact, over the past 15 years, reflection has suffered from a loss of meaning. In becoming everything to everybody, it has lost its ability to be seen." [221] Before going on to examine a reflection rooted in Dewey the author raises the point that defining the term is challenging because it is unseen and vague in nature (we are talking about an individual's thought process), for example how is it different from other types of thought? Further because it is vaguely defined how do we look for evidence of its presence and use and how the effectiveness of the reflection be measured. As educators the last part may be required and effectively a challenge. Is reflection the following of a process and behaviours, such as keeping a journal or discussion with a tutor? By drawing on the work of Dewey the author presents four criteria and discusses them in turn [221]. The four criteria for reflection (with some abridgement) are:

"1. Reflection is a meaning-making process ...
2. Reflection is a systematic, rigorous, disciplined way of thinking ...
3. Reflection needs to happen in community, in interaction with others.
4. Reflection requires attitudes that value the personal and intellectual growth of oneself and of others."

It is worth pausing here and pondering whether these criteria were in the definition you were considering before reading them.

Rodgers distills and discusses Dewey's work to produce a staged framework for the process of reflective thinking. These stages (adapted from [221]) are:

1. Presence to experience,
2. Description of experience,
3. Analysis of experience and
4. Intelligent action/experimentation.

We can see similarity to the Kolb learning cycle here [150] which we discussed in Section 2.4. With the four stages listed here, mapping to the Kolb cycle starting at 'concrete experience'.

Looking at the points in turn we can see that there has to be some sort of identifiable experience. For an engineering student this could be a computer program that they are using that is not working as expected. The student then needs to be able to articulate what the problem is. This provides clarity on the problem at hand. In the case of the computer program it moves from, 'it just does not work', to more of an examination of in what specific way it does not work. The problem then needs to be analysed. Here a cursory or too swift an analysis could lead to not finding the programming error, because the final phase requires the implementation of the actions to solving the problem. Analysis that is too limited may overlook the problem, forcing a return the earlier stages of the reflection process. Once the first three stages have been done the experimentation to the solution can proceed. The most likely reason, from the analysis phase, for the program failure being experimented upon first. If that does not succeed then the other possible solutions could be tried. For most engineering students, this debugging process of code will be familiar from introductory programming courses, but it serves as a good example of the solving of the problem and how experience of solving such problems can make future debugging often quicker as there is familiarity with the problems encountered. Anyone who has taught introductory programming courses will be familiar with the evolution of student's debugging ability of their programs.

This framework is more rigorous than just having reflection as casual thinking about something. The stages provide a clarity to the process, starting with having an experience. Once this process is used, and for an expert this could be done significantly faster than for a student, the outcome from the process could be drawn upon to solve a similar problem. So develops expertise in the area where the experience has been encountered. Over time this ability to reflect or understand how to deal with an experience can become tacit knowledge of the expert.

The use of reflection in engineering is becoming something of increasing interest in engineering education. The growth of reflection in engineering education in Russia reflection is reported [35] as being recognized as important for engineers and in the USA a Consortium to Promote Reflection in Engineering Education (CPREE) has been created and their website includes tools for educators [70]. At the time of writing there were 12 partner institutions that had descriptions of their reflection activities, named as 'field guides'

[69]. These guides give clear examples of how reflection can be used across the curriculum in engineering. Sequential steps for introducing the reflection activities are given, as well as 'tips and inspiration' from the educator who provided the reflection activity. For example, from Stanford University details of how reflective questions are posed to engineering students at the start of an engineering course to encourage them to think about their goals for this course and responses are collected by clicker technology [69]. The results are shared with the class and the instructor encourages the students to apply more metacognition to the course than just simply learning content. Another example from the CPREE 'Field Guide' is from Clarkson University where a skills survey is used to encourage students at the start of an open design course to reflect on their own abilities and to use that in helping with the allocation of tasks and roles [69]. This exercise follows a presentation by the instructor on teamwork and the importance of diversity. These examples provided on the CPREE website are rich and practical examples of the variety of ways that an instructor can encourage a student to reflect. Instructors wanting ideas on how to incorporate reflection into their own work will find the examples a good base for initiating and developing their own ideas.

Guidelines for including reflection into engineering has been presented by Turns et al [267]. They propose the idea of the use of a framework for guiding the reflection. The stages of the framework are: experience; features; lens; meaning; action; intentional and dialectical. Let us look each one of these in turn as part of this framework [267].

Experience – this is the required starting point of the reflection.

Features – the person doing the reflection (the reflector) then needs to focus on some element or feature of the experience that they want to reflect upon. Of course this has to be a feature they are aware of. There could be features and elements in the experience that they may not be aware of that will require a coach or further experience to reveal.

Lens – this is the knowledge that the reflector will use to interpret the meaning in the reflection. For example, if the reflection is on an engineering problem and the feature being considered is a particular part of the problem, then the lens will be the engineering knowledge of the reflector. If it is about the time to do a particular task the lens could be a personal organization and productivity lens.

Meaning – From the reflection after using the lens the reflection can lead to different understandings about the experience under consideration.

Action – After the meaning there can then be a decision about actions that can be taken. For example, with an engineering problem this could be changed to a design, troubleshooting approaches to gather more information.

Intentional – The framework here highlights the need for the reflection to be intentional and not an unconsidered decision on the experience. Without intentional reflection the meaning drawn from the experience can fall into 'jumping to a conclusion'.

Dialectical – Sometimes to understand an experience the use of one lens has to be changed and this tension between the changing perspectives can become called dialectical.

Approaches to using reflection in engineering courses are discussed by Turns, Sattler and Kolmos in a paper that outlines a workshop that they facilitated [266]. In this proceedings paper they raised the point that introducing a reflection activity into a course does not always result in a successful reflection. Indeed, in my own teaching I have found that when including reflection exercises on project work, some students would not reflect and instead would provide a description of what they had done. It should also be said that other students do rise to the challenge of reflecting when encouraged to do so. The reasons for the disparity could be due to a range of reasons, but it probably can be related to either the lack of understanding of what a reflection is, or a lack of comfort in disclosing personal views. The workshop described in the proceedings of the conference provided a staged look at reflection integration, including examining reflection activities, how to deploy them and how to assess them.

In 2015 a paper *Case for reflection in engineering education- and an alternative* [185] was presented at the IEEE Frontiers in Education Conference drawing together the thoughts of three authors who collectively have many years experience and thinking about engineering education and the role of reflection and its use. The paper provides a valuable collection of background and thoughts on the use of reflection, specifically applied to the field of engineering. As the paper indicates the perspectives of a reflective practitioner, a 'staunch disciple of Dewey' and third author who brings an 'alternative commentary', provides the rich discussion on the this subject. Two of the authors bring forward ideas about reflection from early proponents, like Dewey, as well as practical advice; like the need for more data on showing (or not) the degree of learning and development through reflection. The authors note they see an important need in the growth of 'key professional abilities', more than in the past fifty years. They go on to note:

> "In the interim, however, reflective practice has progressed in many professional areas. Hence any question asking what an engineering educator should be doing now in regard to promoting and using reflection should have a revised and enhanced focus, appropriate to this decade of this century." [185]

The third author also provides their perspective on reflection, which they believe is a synonym for 'thinking'. They point out that reflection can cover a spectrum of questions. From the larger questions about oneself and ones' purpose, through to the more pragmatic approach of methods in engineering, which is probably the part of the spectrum that most engineering educators are focusing their student reflections.

Other examinations of the reflection in engineering have been reported. There has be research on the use of reflection in design [4], which connects to the work of Schön.

To draw together some points about reflection that can come from the research described here and earlier in the chapter we can say that reflection is focused thinking about past experiences to help with future experiences. Methods in engineering can be part of the reflection and if we are to consider the effectiveness of reflective thinking then researchers need to have more data collected on the reflective practices in courses.

2.9 Linking the Theories to Engineering Education

Having covered a number of theoretical ideas with regards to education it is perhaps worthwhile to pause here and consider how these can be used in experiential learning in engineering. The aim of this section is to give some suggestions and ideas to the use of the theory and are not intended to be prescriptive and not to restrict creativity in the design of courses and learning activities.

Taxonomies help with designing programs, courses and assessments. Choose an appropriate taxonomy that fits your own teaching style. Use it to set objectives and approaches to the course.

The learning theories and ideas from Kolb and Schön give an idea of how to set up the learning and the assessment. This includes the design and the execution. Design in the way the learning exercises and the assessments (formative and summative) can be formed and released. Execution as in the way the students are guided and encouraged.

The learning styles give one a framework to consider how different students are reacting and handling the learning tasks. Provides a lens to look at the different responses and the ways to encourage and guide.

The concept of deep, surface and strategic learning can also provide insight into the way students will respond to your course and its elements. For some students experiential learning may be a significant departure from what they are familiar with. Sometimes the challenge of adapting to the new approach, a different assessment methods or working in teams may cause anxiety.

Early on we have considered taxonomies and considered the classic Bloom's taxonomy which establishes a hierarchy of learning. The SOLO taxonomy also applies a development hierarchy but looks at the connections and relationships of learning elements. The departure from a hierarchy is Finks taxonomy which is the newest of the three taxonomies. The use of taxonomies is often done at a program or a course level. Elements within a course can fit into the stages of the taxonomy. In choosing the taxonomy there will need to be a consideration of the stages intended in the evolution of the course and its objectives. Whether the course is classroom based, involves laboratories or is project centred can affect the choice of the taxonomy. A more traditional classroom-based course with laboratories could be centred around Bloom's taxonomy. Whereas a project-based course may find that Fink's is appropriate as the various elements such as knowledge, application and human dimensions all come together in the project and the learning of the various aspects are significantly important.

Following an adoption of a taxonomy there can be a guided approach to a course design, with the objectives, delivery methods and assessment (including rubrics) being established. Some of the principles of Kolb and Schön can now be considered in an experiential context. The Kolb learning cycle can help establish stages in the student's learning and processing. Pathways to encouraging reflection can then be considered. Schön's ideas of how to encourage reflection can help the instructor understand how to encourage the learning of students through the reflective discussions. Awareness of learning styles and student learning approaches (*deep*, *surface* and *strategic*) can also be considered in the design of content delivery approaches and assessments.

So, from design through to delivery and assessment the theories raised here can be used to enhance the benefit of the experiential learning of the students. The theory is not to dictate the approach to course design or redesign but is to help guide the instructor in considering the elements within the course. When the course is underway the ideas of learning styles and student approaches to learning can be considered during the running of the course and the operation of the experiential learning.

2.10 Chapter Summary

In this long chapter we have discussed some of the underlying theories that link to education in both general and with experiential learning in particular. This helps prepare the way for some of the details that are to be discussed in the subsequent chapters of this book and in the references cited.

We have looked at key theoretical elements of taxonomies, frameworks for learning, models of experiential learning and learning styles. Depending on your interest and role in experiential learning some of the theoretical work will be more useful than others. For example if you are a course designer or a course leader, the taxonomies will be of interest to you. If you are an instructor then the models for experiential learning, the learning styles and approaches to learning will be valuable to know about and understand. If you are an engineering education researcher then the frameworks for learning will help to understand how a research lens can be focused onto particular aspects of the work. If much of the material and ideas are new to you then do not feel you have to know or use all the theoretical aspects. Some theory may fit better to your ideas and philosophy of teaching and the type of work you are doing. However, using some to be more guided in the approaches you take and the learning that is going on within your course can produce insight and understanding as to what is or is not helping with your students' learning. Like your students, it will be valuable to reflect on your teaching and what you see in your student's learning.

In this book there will be a focus on the situated learning framework and the Kolb learning cycle. These two theoretical approaches provide helpful ways to consider how learning can occur in an experiential learning environment. The situated learning framework can allow the consideration of actions and interactions what the mediating objects are and how the identity of the student as an engineer can be developed. The Kolb learning cycle can provide a way of understanding how the actions and activities in the experiential learning process can lead to gain in knowledge and understanding.

In the upcoming chapters we will look at specific aspects of engineering education that involve and connect to experiential learning, starting with the next chapter by looking at a commonly encountered experiential learning environment in engineering education, the laboratory.

3

Laboratories

3.1 Introduction

The fundamental purpose of engineering is to solve a problem through a devised and constructed solution. Perhaps the earliest way engineering students gain experience of construction type work is in the laboratory. Here theory meets practicality. The test of what was learned in the classroom, from a textbook or other source can be examined, measured and validated. Besides the the application and examination of the theory, there is also the learning of how to use equipment correctly and effectively, often within an unfamiliar and complex system. Problems may be encountered and need to be resolved with an experimental setup. There is the learning of how to analyze data, understand experimental errors and ultimately to report and establish a conclusion of the experiment. This could be validating a theory, examining an effect or determining a level of performance for a component or system. The work could be conducted in a small group of with a pair of students and so there will be the understanding of working effectively in groups. The satisfaction of building something that works to a desired level should not be underestimated, and perhaps is a motivation for all engineers. Others have detailed similar purposes for the engineering laboratory, such as Wankat and Oreovicz [279]. They list the goals of the laboratory can include [279]:

- Developing experimental skills
- Experience of working in the real-world environment
- Building and testing objects
- Discovery of results
- Working with equipment
- Motivation of the student
- Developing teamwork skills

DOI: 10.1201/9781003007159-3

- Networking with other
- Communication skills development
- Independent learning

If we consider a situative learning framework, we can see there are many elements here that can connect to the three areas of action and interaction, mediation and identity.

The form of laboratory experiments can vary due to the discipline of engineering, the subject being studied and the pedagogic aim of the laboratory. This effects the equipment being used, the laboratory environment and the method of guiding the student through the experiment. Traditionally laboratories have been physical spaces with equipment, but with the development of computing power and internet access, there is the possibility of virtual and internet-based, or remote, laboratories. These can have the advantage of cost and space savings, along with safety and the ability for students to access them outside of scheduled class time. The accuracy of the simulation that replaces the real equipment and working environment then becomes the limitation on the accuracy and efficacy of the simulation. An argument for the usefulness of simulations is the known benefit to pilots of aircraft flight simulators, where regular or challenging situations can be explored in the safe environment of a realistic simulation.

Laboratory work is a common and expected feature in engineering programs. Laboratory facilities, the associated equipment and support personnel can amount to a significant investment and running cost for an engineering faculty and department. The quality of the lab facilities can be used to attract people to come and study at the department and many prospective students will be shown the laboratory facilities in visits and promotional material. The lab experience can be such an integral and assumed part of the engineering curriculum that its reason for inclusion may not even be challenged and reconsidered. Despite this acceptance of laboratories as an obvious part of engineering programs, there has been an examination of the reasons why faculty have for including them in the curriculum [68].

What about the content of the laboratory experiments and how does it relate to experiential learning? As was mentioned, the experiment could be about showing a concept which was covered in a prior class. Figure 3.1 can show how this could be related to the Kolb learning cycle. The experience leads around a series of steps resulting in some evaluation on how the experimental result matches the theory and sets the stage for further ideas. Alternatively the experiment could involve problems with the equipment. Any issues encountered here will make attaining the goal of successfully completing the experiment, as well as potentially hinder the matching of the experimental results to theory. This may cause frustration for students as the perceived goal of completing the experiment is being hindered, whereas the issues being encountered may develop the problem-solving skills of the student. If an

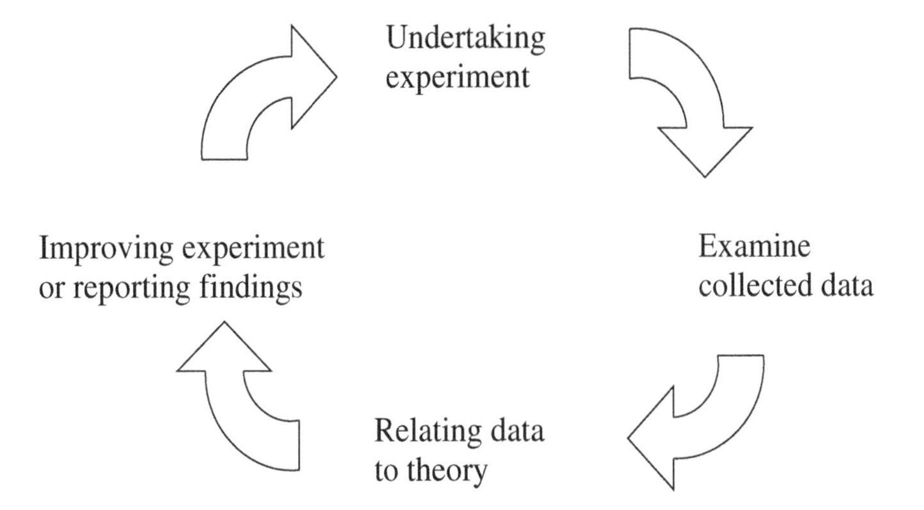

FIGURE 3.1
Kolb's learning cycle applied to a laboratory experiment.

instructor does not tell the students that problem-solving is a valid objective in of itself, then the frustration of the students' could make them feel the experiment is a failure and they could incorrectly surmise they have gained little learning. To illustrate the experience with the problem solving Figure 3.2 shows a suggested cycle when problems are encountered with the experimental equipment and problem solving is employed by the student. Most engineers will relate to solving problems with equipment or a test setup. The encountering of problems in laboratories becomes a valid learning ground for problem solving, although a student who is short on time and focused on getting a good grade may not appreciate the problem at the time.

In this chapter we will look at traditional laboratory experience and the online and simulated laboratory. This chapter was started prior to the COVID pandemic which caused a major shift in educational delivery around the world, moving from in-person to online delivery. At the initial time of compiling the later section in this chapter on online learning the ideas and use of remote access or recorded laboratories were unfamiliar to many faculty. The pandemic changed that and it is expected that there will be a continued examination and publication of results from those studies on online laboratories used during the global pandemic

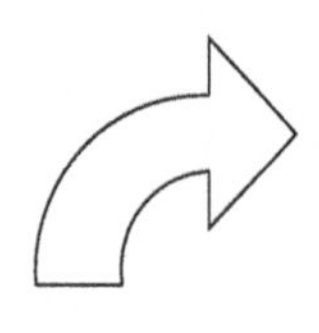
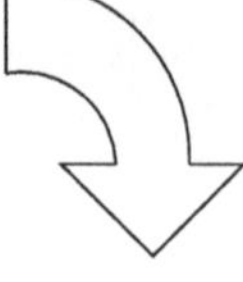
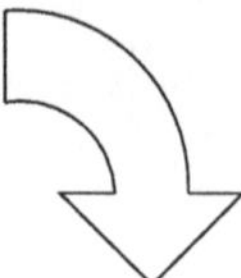

FIGURE 3.2
Kolb's learning cycle applied to solving a problem with a laboratory experiment.

3.2 The Role of Laboratories

It is clear that the laboratories are an expected and important part of the education of an engineer but how did the current form of the laboratory in engineering come about? Feisel and Rosa have reported on the role of the laboratory in the US engineering education system and within their work show there have been changes in the emphasis and role of the laboratory over time, starting with the origins of engineering programs in the US Military Academy at West Point at the start of the 19th century to train military engineers [93]. In these early days the practical aspects were emphasized and complemented by mathematical theory. As other civilian schools in the US developed there was a continued emphasis on practical aspects central to programs. This continued through World War II. A change was initiated with the publication of the 'Report of the Committee on Evaluation of Engineering Education' [116], also known as the Grinter Report, commissioned by the American Society for Engineering Education in 1955. This report recommended engineering education adjust to ensure it kept pace with the rapid developments in science and technology. In discussing laboratories the Grinter Report says

> "The laboratory is the means of teaching the experimental method. It should give the student the opportunity to observe phenomena and seek explanations, to test theories and note contradictions, to devise experiments which will yield essential data, and to interpret results. Therefore, laboratories should be used where and only where these

> aims are being sought. The value of a set number of stereotyped experiments is questionable. The development of a smaller number of appropriate experimental problems by the students themselves under effective guidance will have much greater educational value. The art of measurement—including analysis of accuracy, precision, and errors—and the appreciation of the degree of accuracy economically justified, together with some understanding of statistical methods, are essential elements of laboratory experience. Laboratory reports, when restricted to a few per semester, present a major opportunity to develop skill in the written presentation of engineering information. Stereotyped reports are valueless in teaching the art of communication." [116] Quotation reprinted with kind permission of the ASEE.

It is interesting to note the statement "The value of a set number of stereotyped experiments is questionable" along with "Stereotyped reports are valueless in teaching the art of communication." The use of 'stereotyped' implies a generalized, unoriginal or repetitive laboratory experience, which starts with the experiment and goes through to the communication of the results. The report is emphasizing the need for fewer repetitive experiments and instead increasing the experience of experiments.

Along with a de-emphasis on the 'stereotyped' experiment the report emphasized the increase of basic science. Given the timing, ten years after the end of World War II which accelerated technological development, including the use of the first atomic weapon, and the soon to start race in space exploration, it can be viewed that the report is trying to pave the path for future US engineers to be ready for new technologies. Feisel and Rosa claim the effect of the report was schools "began graduating engineers who were steeped in theory but poor in practice" [93].

Later in 1968 another report on The Goals of Engineering Education by the ASEE commented [275]

> "practices continue unquestioned such as the 50-minute, three-day-a-week lecture, the chalk-board as the main visual aid, the one-teacher course or section, the two- or three-hour laboratory, ten-year-old cook book experiments, the 18-20 hour load, the 16-week semester," [275]

Again there was a noted concern with the laboratories and a 'cookbook' approach. Despite the Grinter report, 13 years earlier, there was still a perceived problem with a laboratory that is formulaic in approach and that does not change. Long after the Grinter report the question can be raised today how many laboratories in engineering schools currently are recipe-like?

Why then was (and perhaps still is) an inertia to changing from the so-called 'cookbook' form? In 1971 a survey was made of 20 UK university Electrical Engineering departments to examine the changes being made in laboratories, especially at the first year level [159]. All departments replied except one and the responses indicated that at only one-third was making "a real move" toward changes, usually involving more open-ended and project-like laboratories. The reason gleaned from the survey for not changing things was given as space, financial constraint, facilities for making items and

> "...finally but by no means least, the volume of extra work which this type of practical work inevitably puts on members of staff apparently without the compensating features of promotion or career prospects." [159]

Now, 50 years later these reasons may still be valid. There may also be the added factor that the laboratories, although not perfect, had served well in the past and most professors when students would have learned with similar styles of laboratories. So, the inertia to not change because of the time, cost, resources and lack of reward is reinforced by a perception that it has served suitably in the past.

Similar discussions about changes to laboratories will no doubt have been discussed in other science-based disciplines. One interesting series of discussions at a 1968 summer workshop on introductory laboratory in physics was observed and documented by two education researchers (who were not physicists) [138]. The 'outsiders' impartial observations and insights of the discussion makes for interesting insight into the differences of opinions on changes to the laboratory. Similar to engineering the concern was expressed at the meeting over the cookbook laboratory experience of students and observations on the opinions and attitudes to them by physics instructors were documented. The practical and logistical benefit of having a relatively large number of first year students do the cookbook form laboratories was counterpointed by the dissatisfaction by the lab experience being somewhat unrealistic and unlike 'modern experimental physics'. The observers of the discussion describe the two extreme opposites of the opinion divide, on the one side those that want to cover as much of the fundamental knowledge of the course content, and on the other those that want to remove constraints to allow students to learn by their own doing and have some fun. The authors grouped three areas of issues that were discussed

- student motivation
- aims and benefits of the laboratory
- constraints on divergence and open-ended nature of the laboratory.

The paper recommends a divergent laboratory which somehow tries to balance structure and openness, with some guidance and instruction given, perhaps early on and then later with time for students to explore their own aspects of the laboratory and interest [138].

The disagreement on the role of the laboratory as displayed in that physics meeting in the past may highlight another reason for the lack of change from the 'recipe' format and that is the difference of opinion by faculty and instructors on the laboratory role. Regardless of the constrains, whether logistical, fiscal or pedagogical there has been a slow move away from the prescriptive laboratory experience in engineering toward more open and even project-based laboratories.

The Grinter Report in the 1950s encouraged the development of more basic science into engineering, the last few decades of the 20th century and beyond saw an increasing emphasis on the inclusion of design, as well as an adoption of the more open nature to problem solving that design problems bring. Details of how one mechanical engineering department approached this is documented by Incropera and Fox [134]. The inclusion of design and more open-ended problem solving developed more opportunities for investigative work, team working and even involving industry. In [134] there are details of small (about six weeks) design problems in the laboratory of the heat transfer course and the author's report:

> "these projects are involving a combination of *analyze, design, build, measure,* and *explain* activities, with students fabricating equipment in the School's Student Shop and measuring performance using re-sources of the Heat Transfer Instructional Laboratory." [134]

So now the move to including more design aspects into programs gives a direction for the modification of some laboratories away from the cookbook approach. Though as Incropera and Fox mention in their examination of their revising of the curriculum there was some resistance to change and consideration that there was a "softening" of the curriculum rather than "enriching" [134].

The movement from the 'cookbook' procedure-based laboratories, to a more open and design-orientated form, shows that there can been a blurring between the conventional project form and laboratories. This allows more opportunity to include features that may have resided more traditionally in a later stage project course, such as planning, communication (formal and interpersonal), testing and design work. This creates an environment that is closer to real engineering work, compared to the step-by-step laboratory. Bidanda and Billo reported on a course that became a project to develop automatic data collection laboratory experiments for student use [29]. Working in groups the students acquired equipment and developed the labs, reporting along the way. At the end a student survey revealed that the students found the work very realistic (from unboxing equipment to developing the lab), they rated highly their depth of understanding in the area they were working with, but the breadth of knowledge was rated lower. Here there may be the potential downside for more open project-like laboratories, compared to a series of structured laboratories that cut across a range of topics in a course. There may need to be a balance between the depth and breadth, as the benefits of the project-like structure may excite and inspire students about engineering, which outweighs the lack of breadth.

The objectives for the use of laboratory experiments can vary dependent on the course but there are some identified reasons that a laboratory can be used. Wanakat and Oreovicz provide reasons in their book Teaching Engineering [279], p. 180, and those were listed in the introduction to this chapter (see 3.1). Perhaps the most common laboratory objectives, from that list are 'experimental skills' and 'discovery of results', with the communication coming in as a strong secondary reason. 'Building and testing' along with the 'working of equipment' are possibly assumed to be implicit in the lab exercise. Perhaps a less considered reason from the list is 'motivation of the student'. Most engineering students and faculty enjoy working in a laboratory, but the role of motivation may often not be explicitly written into a laboratory's objectives. It is worth reflecting back to some of the educational theory here and considering a situative framework to examine laboratories. Unlike a lecture on an engineering subject, a laboratory provides a more authentic engineering setting to learn. There are items to construct and investigate. There is equipment that needs to be used correctly, so as to make measurements that need to be processed. This authenticity can be a motivator for many. It also becomes a learning environment with the equipment and some of the specific laboratory language being artefacts that can mediate learning [140]. There will be interactions with instructors, teaching assistants and laboratory technicians, that will be providers of assistance and explanations. The student is the learner in this environment, and depending on their past they have a variable level of knowledge of the equipment and laboratory. Here perhaps the concerns that were raised in the previously mentioned reports were drawing attention to the value of the student needing to work within the laboratory, to gain maximum benefit and understand their current position within it. Providing the 'cookbook' recipe-like approach reduces or even removes some of the learning. The open nature of the experiment is reduced to a list of instructions on what to do, though this style maybe efficient for the instructor, teaching assistants and others. That said though, perhaps students are wanting to have success in the laboratory, similar to the experience of solving a problem in a text book. Following the steps correctly results in an expected and correct answer. However, if an inexperienced student follows the steps in an experiment and still does not get a correct answer, perhaps due to a misinterpretation, there is a risk this could affect their motivation and potentially undermine the student's trust in the process of the laboratory. Typically, mistakes and missteps should encourage seeking out help from others. We will return to the discussion of the recipe laboratory approach in a moment.

So what are the primary reasons for laboratory experiments? Are they designed for illustrating concepts and supporting the class room learning, or are they designed for the experience of doing experiments? In a 2017 study on faculty decisions on laboratory integration in engineering education [68] the authors comment that

> "In a recent systematic literature review, Coutinho [17], found that 22 out of 23 articles describing laboratory activities reported 'the development of conceptual understanding' as the main learning objec-

tive, while only a few articles mentioned different learning objectives such as developing instrumentation and design skills."

The cited literature review in this quote is unfortunately an internal document and unpublished.

It is not surprising that the focus is on the conceptual understanding however, there may still be a strong development of equipment-focused skills implicit within those laboratories. Most readers who are engineers or scientists will recall using challenging equipment in their undergraduate courses to measure some concept, parameter or constant. Perhaps those memories of that laboratory work are stronger than many lectures that were attended. Coutinho et al report that 60% of the literature they reviewed made use of the 'cookbook' style approach, which they also mention has "limited pedagogical efficiency" for engineering undergraduates [68]. Again, this mention of the 'cookbook approach' to laboratories reappears, indicating it is potentially common practice and identified as not the best pedagogic approach. Looking at the advantages of this approach, it allows detailed and safe instructions on the setting up and conducting of the experiment. As well, it will provide experience in reading, understanding and following instructions. However, it could reduce the amount of actual thinking involved in constructing and conducting the experiment as well as closing off different approaches and open-ended experiment design. Abilities that are often desired to be encouraged in engineering. If an experiment were to become less prescriptive in the process, as well more "developing instrumentation and design skills," as mentioned in the previous quote, then perhaps the experiment will move more towards being a mini-project. Certainly projects in engineering education are considered to be more open in nature and allowing more design to be included in the project. We will discuss projects in a later chapter but it is interesting to explore the borders between laboratory experiments and projects that form the main part of the practical work in engineering programs.

3.3 Student's Perspectives of Laboratories

We have looked at the role of the laboratory from a curriculum perspective, but before we go deeper into physical and virtual laboratories what about the perspective of the students? They are the ones who undertake the laboratory activities. Often these activities have been established years prior and perhaps not by their current instructor. There can be a significant amount of time spent in the laboratory environment in a term. Likely each laboratory undertaken contributes to the final grade for the course and so the importance of the time doing the experimental work becomes important. Student's questions in the laboratory room are often dealt with by teaching assistants who may also be involved in marking their work.

One study done by Lin and Tsai looked at their preferences for the learning environments of a classroom and a laboratory. Their study surveyed over 300 volunteering students and they completed a questionnaire called a Conceptions of Learning Engineering [160] developed by the authors. The study split the stages of learning into seven areas: i) memorizing, ii) testing, iii) calculating and practicing, iv) increasing one's knowledge, v) applying, vi) understanding and vii) seeing in a new way. These stages have similarities to Bloom's taxonomy [36] as discussed in Chapter 2. The questionnaire posed 39 Likert scale questions to the students to assess the stages the students related to learning engineering. Example questions were "Learning engineering is to prepare answers to the questions given in an exam" (Testing stage) and "Learning engineering is to understand the mechanism for the procedure of experimentation" (Understanding stage). Subsequently, students were asked to rate their learning environments by allocating 100 points between a classroom environment or a laboratory environment. Finally they were asked to complete two short essay questions "In your preferred learning setting, what do you understand by learning engineering?," and "In your preferred learning setting, how do you learn engineering knowledge?". The results showed three clusters of preferences for the learning environment, 54% for classroom and laboratory, 24% for classroom and 22% for laboratory. These groups showed some different views on what learning is; those favouring the classroom environment tended to prefer *testing* and *calculating and practicing* whereas those favouring the laboratory tended towards the stages *increasing one's knowledge, applying, understanding* and *seeing in a new way*, which are tending towards the higher levels of Bloom's taxonomy. From the data including the essays the authors concluded there is a transition towards the higher-level stages of learning as there is a change from favouring the classroom to the laboratory. The results also showed a perceptual change in the role of the instructor, from a lecturer in the classroom environment to more of a tutor who provides support. The laboratory also provides more of an environment to learn from peers and also a place where they can conduct more of a meaning for their engineering knowledge.

This study raises an interesting point, the possibility that as students develop more of an understanding that learning about engineering involves self development of knowledge, applying knowledge and looking at new ways, there is then an awareness that the laboratory environment is the preferred place to learn. Now, whether the laboratory environment stimulates the higher levels of learning, or the awareness of learning at higher levels stimulates interest in laboratory work cannot be separated here. Likely there may be some of both and a feeding back and forth. If we consider a situative learning framework (Chapter 2) it may be that the environment, the equipment, the experimental work, coupled with discussions with instructors, teaching assistants and peers, can all be factors that make the student feel that they are moving further towards becoming an engineer and learning more about what is involved in engineering. Whereas a classroom environment is similar to what they have encountered since starting school, with only the material covered by the instructor getting more complex. It also stresses the importance of developing the laboratory experience for students, perhaps more so for those who are slower to rise to the higher levels of approaching learning. The movement from the learning from lectures in a classroom to the laboratory mirrors the type of learning at the post-graduate level in engineering, where there are fewer classes compared to undergraduate work and the final thesis work involves time spent doing research in a laboratory.

3.4 Traditional Laboratories

It is typical for engineering courses to have an associated laboratory component. The laboratory is to illustrate or reinforce some of the principles and ideas that have been introduced and discussed in the course. Some courses may be totally laboratory based. As mentioned earlier in Section 3.1 the aim of the laboratory can be multiple, from developing experimental skills to teamwork, but often the motivating force will be the using of professional equipment, encountering the challenges in a construction or measurement and learning how to interpret results. Laboratories can be the construction and testing an electrical circuit, the writing of code to undertake a certain task or the measurement of a behaviour in a fluid flow lab. Each field has its own areas and equipment that students need to become familiar with the undertake future work. More than a classroom the laboratory can become a place for situative learning. Here the knowledge and expertise with equipment, tools and the materials being used, coupled with working with instructors, technicians and other students can show the progress of understanding and development of a student. In a classroom lecture there is little apparent difference between graduate students and first year students as they sit in chairs and listen to the instructor, without looking at material being covered. However, in a laboratory, the ability and the level of the student can become apparent from observing how a student is proceeding and interacting with equipment and other. The situation can reveal ability.

We have already noted in Section 3.2 some of the long-standing criticisms of 'cookbook' format laboratory descriptions. In this section we will look at some attempts to overcome the challenge of repeating that form of laboratory. Although not exhaustive this will help show approaches that can develop the learning in the laboratory experience.

Combining the Kolb learning cycle with the laboratory experience Abdulwahed and Nagy have reported on the how steps to enhance the prehension dimension of the cycle (the link between concrete experience and abstract conceptualization) [1]. They reported on an attempt to overcome the rote following of a laboratory procedure by the inclusion of virtual and remote laboratory sessions to prepare for a 'hands-on' process control laboratory in chemical engineering. The advantage of the additional remote processes allowed students in their own time to conduct and if desired repeat the experiment in the virtual or remote form. The aim of this virtual pre-lab was to introduce more preparation (beyond just the lab manual) and to potentially encourage reflection, which is a step in the Kolb cycle for learning. Using a control group the authors found through post-lab testing that those that prepared with the virtual or remote lab had test scores that indicated more in-depth learning. The authors conclude that their work gives evidence of better transfer of knowledge along the prehension axis, from a laboratory experience to a conceptual understanding. The use of a post-lab test was considered to give the students a chance to reflect on the experiment enhancing the Reflective Observation stage of the Kolb Cycle model that the authors were using. Interestingly when both the control and experimental groups were asked the question at the end of the course "Would you like the idea of conducting post-lab real experimentation through the Internet (i.e., from your home PC) after the lab for enhancing your report or testing further ideas?" [1], the response was more favourable from students of the experimental than the control group. So indicating an acknowledgement of the benefit of the virtual labs by the students who experienced them. This experiment showed the benefit of the extra preparation time. This does bring up the point of the use of the student's time and the amount of time that can be spent on laboratory work. If extra laboratory preparation has to be done then there needs to be a careful examination of the time students spend on laboratory-related work. Just adding extra pre-lab work onto students for every lab in a suite of labs will lead to further loading of students, who are probably already short of spare time in their course load. However, the careful adaption of the laboratory content and schedule may provide a clear benefit in the learning and understanding of the material in the laboratory.

Another development in laboratory education was influenced by the Kolb learning cycle was reported by Chen, Shah and Brechtelsbauer [63]. The detail the development of a chemical engineering laboratory for a cohort of about 100 third year students, called the Discovery Laboratory, at Imperial College London. This lab was within the four-year integrated Masters program and follows two earlier labs, the Foundation Lab and the Knowledge Lab which were in the first and second years of the program, respectively. Following the sequence of stage laboratory development the Discovery Lab gave the third-year student more independence to investigate a particularly open-ended problem relating to chemical, industrial and pharmacological process. The report list 11 specific areas/project. Allocation of the projects is done to groups of four students and is based on their expressed interests and the students were allowed to redefine the project too. Completion of the project work was done in seven weeks, though advanced notification prior to the start of term allowed students the opportunity to undertake some preliminary studying. Once the course was started there was an outlined schedule of stages, such as literature searches, risk assessments, experimental work and reporting. The splitting the course into stages of preparation, execution and reporting, was intended to follow the Kolb experiential learning cycle and the experimental stages was to allow the students to go through the cycle a few times. The staging also followed a Design-Analysis-Evaluation process too. Because of the research nature of the laboratory teaching assistants provided training in the use of different equipment, as well as the practicalities of the planned experiments. The academic instructor took the role of subject expert and facilitator. Part of the effectiveness of the course was evaluated on the student feedback through the institutional online survey tool. Comments indicated support for the experience that had an authentic research or industrial-like experience; a course that pushed them further than normal; and the development of the their softer skills.

In looking at the Discovery Lab it can be clearly seen that the openness of the laboratory breaks well away from the 'cookbook' format. However, there was structure provided through organized staging of the process and the use of the TA and instructors to guide the students through the choice of experiments that they wanted to undertake. This could make the course feel more like a research or industrial approach to work, which the students likely recognized as valuable. The use of a specific topic for the project could be seen as limiting and sacrificing exposure to a broader range of labs and topics, which a 'cookbook' approach may provide. However, the value of the student-designed investigations, experiment design and reporting would more than compensate for the breadth, especially if the students learned more about the process of conducting laboratory investigations as opposed to following step-by-step instructions and reporting on how the results fit theory.

3.4.1 Laboratory space and makerspaces

In recent decades work spaces have been changing for employees in corporations, with the provision of working environments that can inspire collaboration, innovation and creativity [110]. Similarly, there have been developments in learning spaces and classroom design [203], which will be discussed in the next chapter. Laboratory spaces have also been undergoing some changes in recent years, particularly with spaces for collaborative design and construction, known as makerspaces [23, 103, 281]. These academic spaces have sometimes followed ideas of hackerspaces (later often known as makerspaces) that have been spaces created outside of universities by communities of individual 'hackers' or 'makers', who have grouped together to share space and equipment for the building of projects.

Traditional laboratory spaces need to be regularly updated. Narum in the 1990s gave the usable lifespan of "research and research-training facilities" in science as about 30 years [193]. Laboratory spaces for education can be updated in a series of ways such as the technical equipment within it; the learning environment infrastructure, such as the chairs, tables, displays and computers and then there is the physical infrastructure including the room size, lighting and air condition, etc. The order of the need for upgrading is probably in the order presented here. Individual departments will be upgrading equipment on a regular interval, as well as less regular updates with the benches displays, etc. For the physical infrastructure of a laboratory room an engineering faculty or the university may undertake those changes to meet the needs of the institution. The case study included in Section 3.5 started out with the need to update the efficiency of the ventilation and developed into a pedagogic renewal as well.

Today, access to the laboratory equipment can be possible via the Internet and so it can be important to consider how the student users are accessing the equipment. Users of a lab space may be physically in the room or accessing remotely. In my own department during the COVID-19 pandemic there was access to electrical circuits and measurement equipment via the Internet. Web cameras were mounted so that users could see the circuit and the display and control of equipment like oscilloscopes could be accessed by connecting the equipment to the Internet. When there was a limited return to in person teaching, I would have some project students in the lab space working, but knowing that after their lab time the next group was accessing the laboratory remotely. So it was important that students did not switch off the computers that the equipment was connected to and that the circuits had not been moved outside of the web camera's view. Students, TAs and instructors knew the space was shared by users in the room and remotely. Although, this scenario happened during a pandemic, this could be a more common way of operating as laboratory spaces can provide remote access to users. The idea of laboratories for 'hands-on', remote and virtual use is discussed by Müller and colleagues, who consider the idea of mixed reality learning spaces [191]. In this work they consider a collaborative learning space that has a mixture of real, remote and virtual labs, including research on this area using a computer automatic virtual environment (CAVE). We will discuss remote, recorded and virtual laboratories in the next section.

As mentioned earlier makerspaces have become a feature in more engineering schools. The fundamental idea of the academic makerspace is similar to the community-based makerspaces which is a open space for construction and fabrication. This means the space often has equipment for machining materials, as well as soldering stations for electronics and possibly items like sewing machines if fabric is used, for wearable technology for example. The difference between the the community-based and academic makerspaces is the academic facilities can have funded support from the institution as well as expertise from technicians and faculty. Unlike other departmental shop facilities the aim is generally to have the space more open to students of different departments and for projects that may be linked to courses, extra-curricular work such as through student competition groups and club activities or possibly personal projects. The space becomes a more informal learning space and a centre for sharing knowledge and expertise as well as a space to learn about design and construction. An informal learning space is a place where students can learn outside of a scheduled class. So a formal learning space is a classroom, but a space in a library where friends can congregate to study and discuss coursework is informal [277].

Descriptions and analysis of some US-based makerspaces have been detailed in papers by Wilczynski [281] and Barrett et al [23]. These examinations look at makerspaces at MIT, Stanford, Yale, Georgia Institute of Technology and Arizona State amongst others. Barrett et al provide a comparison of how the makerspace is organized and run [23]. For example, whether faculty or student run and whether there is specific support staff or not. Whereas a traditional machine shop facility or lab in a department would be run by a department with faculty oversight and with technical staff support, the openness of makerspaces can produce quite a different model of organisation. The openness for both access and within the space allows for collaboration. Here is where the makerspaces can have elements of some of the more innovative working environments that have been emerging in recent years [110]. In these spaces creativity is encouraged through innovative work space for employees. As universities create informal learning spaces, like makerspaces or learning commons in libraries, they provide students the opportunity to learn in environments that many progressive companies are trying to create for their workforce.

Wilczynski provides a number of suggestions and guidelines for academic makerspaces from their study [281]:

- "The mission of the academic makerspace must be clearly defined";
- "Ensure that the facility is properly staffed";
- "Open environments promote collaboration";
- "Aligning access times with the student work schedules";
- "Providing user training is essential";
- "Attention must be devoted to establish a maker community on campus".

The types of facilities that 40 different US institutions have in their makerspaces is detailed by [23], who lists whether they have 3D printing, laser cutting, wood shop, metal shop, electronics, textile, computer and white board facilities.

The benefits of such facilities for design work and support for project work, from introductory level to capstone is significant. This type of facility will become not only more popular in future in engineering schools but most likely expected.

3.5 Case Study

Activity: Creation of a large laboratory space.
University: Carleton University, Ottawa, Canada

When faced with a significant renovation to improve the energy efficiency of a building which housed the main undergraduate chemistry laboratories there was an opportunity to renovate the laboratory space as well. There were four main undergraduate labs used for the junior-level chemistry courses (including for engineering students), which all adjoined each other and had space for technicians. The decision was to combine the four labs into one larger space that could house over 100 students at one time and even cope with two different courses being in the room simultaneously. The laboratory could also be equipped with improved equipment, including benches, lighting, fume hoods and audio-visual (AV) equipment.

The resulting large laboratory room could accommodate 120 students working in the space at once. In each corner of the room there was a large screen and associated audio equipment and cameras so that presentations and demonstrations for students could be done. As well, the AV stations could be linked so that projection of the screen could be shown in all four corners of the room if desired. Fume hoods throughout the room had glass on both sides so that visibility was good across the room for students, teaching assistants (TAs) and instructors. The new lighting provided good and even illumination across the whole room. Laptops were available for students to use and could be connected to some of the technical equipment in the lab. Accessibility lab benches were included too. All stations had under bench space for student coats and bags to reduce clutter and associated risks of tripping. Safety stations were strategically positioned around the room, including safety showers in each corner, spill kits and gas shut off controls. The original energy efficiency was also achieved with one of the largest variable air volume controls implemented, as well as heat recovery being added. This helped reduce the building energy consumption to below half of the consumption before the renovations.

The laboratory came into operation in 2008 and became one of a small number of large undergraduate chemistry labs in North America, becoming known as the 'Superlab' on campus. The benefit of the change was a bright, safe and efficient room for students to work in. Students could capture data from the equipment and then copy it for later analysis and report writing. The original data also was retained and stored on site. Moving from four separate rooms to one combined room allowed the TA to student ratio to decrease a little, from 1:12 to 1:18, as the TAs found it easier to move around the room to monitor and assist students. Prof Robert Burk, the chair of the Chemistry department at that time, who requested the renovation of the lab, noted that there seemed to be an unexpected bonus that some TAs were competing to have the best presentations to students. Consequently, the overall quality of TA presentations rose. When asked if the large lab space created a problem in control of noise and movement, he noted the contrary and that the space became surprisingly peaceful, as students worked effectively and efficiently in the laboratory. Seeing so many students working on the same experiments appeared to energize the students. He reported there was no problem with having two different courses running at the same time. There was sufficient space and flexibility in the provisions of the room that first years and second years students in their different courses could work effectively in the same space. There was an added benefit that first years could have an opportunity to see what they may be doing in the following year. Following the success of the junior laboratory renovation, the senior laboratories were also renovated in 2013, though not on the same size scale. With the senior labs there was more advanced technology included, many of which were computer controlled.

One final added feature was the inclusion of chalkboard walls in corridors along two long sides of the lab. These corridors to access the labs are around 100ft long so this made for long chalkboards that many individuals or groups could use. Use of the boards was open and Prof. Burke noted that undergraduates, researchers and faculty would use this informal writing space. He said it became his favourite 'classroom'. Towards the end of a term as examinations were approaching he reported seeing many students pouring out the course content onto the many feet of the chalkboards. He could then walk up and down and discuss and clarify details with the students. He noted this became such a low cost but effective teaching resource for all.

The impact of the 'Superlab' has been significant. Not only with the energy savings in the refurbishment but also in the pedagogic value that was designed into the changes. Sometimes the simplest of changes, such as the addition of a chalkboard in the liminal space of a corridor to the lab, can inspire learning.

Additional information and photographs can be found at https://carleton.ca/chemistry/about/facilities/superlab/

3.6 Remote, Virtual and Simulation-Based Laboratories

In the introduction to this chapter some of the reasons were given for using virtual laboratories, recorded experiments or simulations. They may not seem to offer the same degree of authenticity that working with real components and test equipment offer, but there may be good pedagogic reasons for using virtual laboratories. Before going deeper into these types of experiments let us list some potential advantages and disadvantages.

Advantages

- Students can receive the same data.
- Experiments can be run under different conditions, including some that could be dangerous and/or extreme.
- Overcomes potential fiscal constraints.
- Allows access to the simulation and data from different locations and outside of normal working hours.
- Creates familiarity with simulation tools.
- Easier to change and update.
- Development and use can be shared by multiple institutions.
- Requires less room space and time to set up than physical equipment.

Disadvantages

- Little or no assembly of equipment by students.
- Not using measurement equipment, which can be complex and need familiarity.
- De-emphasis on preparedness. Experiment can be easily restarted to initial conditions. Not always possible to do with a real experiment.
- Student interaction through a user interface and not in the presence of the experiment
- Requires computer, software and possibly Internet access.
- May not accurately mimic reality.

Both physical and virtual laboratories have the potential for discovery-based learning [2] and there could be limitations with both. For example, with virtual experiments a negative could be the constraints due to the user interface, whereas a positive could be the virtual experiment could show how an experiment can move into unsafe conditions. When dealing with physical experiments safety is a key consideration and so limits of operation and safety guidelines need to be adhered to. So, the discovery of the limitations using a virtual experiment could be significant for learning and be an aspect of exploration rather than avoiding for safety in a physical experiment implementation. Awareness of the advantages and disadvantages of the way the experiment is implemented can also help in maximizing the benefit of the learning experience, and adjusting the objectives for a given experiment (whether physical or virtual).

A number of studies have looked at the role of virtual laboratories and student perceptions towards them. Some of the findings that have shown:

- Increases in awareness of experimental design, critical thinking and ambiguity in virtual laboratories over physical ones. Whereas the physical labs have shown an increase in laboratory protocols and specific content over virtual labs [154]. Both physical and virtual labs were an industrially situated scenario. The authors interpreted the differences due to instructional design and not the type of laboratory.

- Simulations can improve conceptual understanding over use of a physical lab and no laboratory. This was done with a basic DC circuit simulation that had virtual measurement instruments and an animated indication of the flow of electrons [98]. The user interface was designed to aid interaction with the simulation.

- The displays of a simulated chemical process were created using Labview™ to resemble a typical industrial process display [280]. The authors concluded the physical (or 'tactile') and virtual forms of laboratory complement each other and the simulation allowed challenging experiments to be undertaken to meet the lesson goals. They reported their students appreciated the simulations but would not want to only have that style of laboratory.

Another example study needs more details, as some of the carefully examined results show interesting findings. The student laboratory used in the study was a measurement of acceleration due to shaking, using a spectrum analyser [161, 162]. This was connected to both the driving circuit and a laser Doppler vibrometer, to determine a transfer function. As the laboratory was computer controlled it could be controlled either in person or remotely. As well, a simulation could be created that had the same computer user interface and so could appear as it was controlling and reporting the information from the laboratory. By using the equipment in the same room, with the same user interface, requiring the same laboratory report to be written and conducting the same post-laboratory questionnaire the authors could look at differences coming from the reports, or the questionnaires when over 100 students undertook the laboratory in one of the three access modes (direct, remote or simulation). The experiment using the different modes was done with [161] and without [162] audio-visual feedback. After careful examination of the student reports and the results of the questionnaires the authors found the access mode:

- had no effect on whether the students found the experiment stimulating (levels were about 85–89%) but the reasons appeared to vary with the access mode,
- affected some of the learning outcomes,
- affected the perception of the laboratory objectives but *not* the perceived outcomes.

There appeared to be a mismatch between the perceived objectives and the perceived outcomes by the students. When asked what access mode they would have been preferred the favoured form was in the presence of the equipment, regardless of the access mode used by the students. The most preferred mode for the remote or the simulation groups was the mode the students had used. Which indicated some familiarity with the alternative modes did make them more acceptable. The authors considered a constructivist approach to what was occurring and postulated that the students knowledge of the mode of the laboratory may lead to a different context of learning, and as a result the different learning outcomes may arise. This is an interesting postulation as those students that know they are looking at simulated data may be focused on the analysis of that data to understand the features being seen and fitting to theory. Whereas students knowing they are working on the physical equipment next to them may miss features in the data, as they consider them to be due to the behaviour or error of the equipment and not meriting further investigation.

So what can we glean from these investigations into simulations and virtual laboratories? It does appear that both are a suitable and effective form of experiment, though students may have a preference for physical experiments, which may be a desire for an 'authentic' engineering experience. This may change over time as students may experience more simulated experiments in their courses, as well as an awareness that many engineers conduct simulations to better understand the physical situation. The virtual experiment can be used to complement physical ones and the learning outcomes may be different, perhaps being more focused on the theory. Which brings us to the objectives of the experiment, which appears to be important and the instructor who is designing the experiment may want to carefully consider the planned objectives and goals, not forgetting the students may perceive different objectives. The virtual experiments do provide an alternative form that allows them to be designed to provide careful learning. Finally, in their conclusions Lindsay and Good point out that the non-physical experiments should be considered as "pedagogical alternatives, rather than merely logistical conveniences" [162].

3.6.1 Remote laboratories

With the extensive development of the Internet it has been possible to create access to experiments online [10]. Control of physical equipment and collection of data can be done remotely from the physical equipment. This helps provide access to equipment where there may be safety concerns, but also allows the sharing of an experiment (and potentially the associated cost) across different institutions and around the world. The MIT initiative iLab created a range of remotely accessible engineering laboratories from microelectronic circuits to heat exchangers, chemical reactors to a weather station [187]. They report access from around the world including China and Africa [187]. The three goals for that project were to create an intuitive infrastructure, scalable architecture and cross-institution cooperation. The benefit of sharing can be considerable to the participating institutions and their students.

Another collection of experiments, with simulations, this time from Europe was the 'Library of Labs' or 'LiLa' initiative. This started in 2009 and when eight European universities and three companies partnered to produce a range of laboratories [214, 257]. The challenge of collecting and making accessible a range of remote and simulated laboratories involves not just the running of the laboratories through the Internet but also the booking and resource allocation. Just like all laboratories there has to be a scheduling with the opening up of access to other institutions the problem of scaling can grow [173]. The design of the laboratory modules was such that they could be used within a learning management system. In one report on LiLa the breadth of the library was 41 remote experiments and 218 simulations (or virtual) experiments. These were distributed across the disciplines of physics (150), chemistry (6), engineering (7), mathematics (75) and computer science (21) [40]. It is interesting to note that there were five times the number of simulation-based laboratories compared to the remote control physical experiments. This may have an indication of the complexity (at that time) to create and administer the access and control of an experiment online, compared to developing a simulation. Some have tried to overcome the access issues, with the aim to produce massive open online laboratories (MOOLs) [95], or laboratories that can complement MOOCs. It is acknowledged that there needs to be some way to implement preparedness to gain access, usually through assessment and guided progression.

The NetLab project in Australia allowed students to remotely build and measure electrical circuits remotely [195, 196]. This is an interesting variation as it allowed the circuits to be constructed, rather than having been pre-built and available for access for measurements. Students from both Australia, Singapore and Sweden used the remote setup to work collaboratively and this was a planned key feature of the work, to have students gain intercultural experience of working in international groups [195]. This now exploits another advantage of having a remote laboratory, as the laboratory now becomes available to many outside the campus and even the country. The interactions between the students from different countries were observed and recorded. The findings showed that the politeness between the students could lead to a lack of directness and ambiguity. Perhaps a commonly encountered situation in most new international professional interactions, so maybe the learning situation also emulates reality. The authors of the study note the balance needed in increasing directness, avoiding misunderstands between collaborators whilst maintaining professional politeness and not misconstruing directness as rudeness [195]. This work shows how an extra layer to the experiential learning can be added as students not only work on an engineering lab but they do it in collaboration with others from a different country.

The ability to control the safety with the laboratory experiment is an aspect that is of key importance with perhaps one of the most ambitious remote laboratories for academic use, is the Internet Reactor Laboratory which uses the PULSAR reactor at North Carolina State University [271]. Through this program students can undertake an experiment with a nuclear reactor using audio-visual links to the control room and the personnel working within it. Access to data from the laboratory is done online. The students can control the experiment, with the safe operation being under the control of the operators in the control room at NC State. The type of experiment that can be undertaken includes normal reactor startup and power operations, approach to criticality and heat balance power calibration [271]. The Internet access to the reactor has allowed students from outside of the US to use the reactor [167]. This scaling of the remote-controlled experiment, now involving people in the control room, not just computer control, shows the potential, the investment and perceived value of having students experience the operation of a system that is not only rare to find but has safety levels to maintain. The benefit to the students is beyond what they could obtain from a visit to the reactor and watching the operations. They are safely in control of the experiment and receiving effectively real-time data from an experiment they undertake.

Although the route of having an experiment online may be considered an alternative environment for an experiment some organizations are using online access to experiments for investigators to run experiments, avoiding the duplication of equipment at different sites, or travel by personnel. One example of this is a remotely accessible hybrid vehicle motor experiment, that was set up to relay real-time data locally and remotely [237, 238]. This shows not only an application of remote laboratory access beyond use for student education, but also the potential for a student laboratory or project that looks at the engineering and experimental impact of the remote control of an experiment.

At the time of writing this the COVID-19 pandemic is underway and the majority of educational institutions around the World have moved to online and distance learning. As engineering schools adapt to the delivery of online course materials, one of the largest challenges has been how to deal with the laboratory element of courses. It is interesting to speculate whether there will be growth in the area of remotely accessed laboratories following the pandemic.

3.6.2 Recorded laboratories

Another form of virtual laboratory is a recorded laboratory. Although this can reduce a degree of interaction with equipment it can allow experience to an experiment that is too large for a laboratory, or is too unsafe for students to be directly involved with the equipment. It does also allow students the exposure to identical situations.

One example of this form of experiment was a gas turbine laboratory created by the US Military Academy [273]. Because of the use of gas turbine engines in the military the army cadets need to become familiar with the principles of the engineering and the related thermodynamics. The objectives of the experiment were to understand the use of turbine engines in the military, to understand the design of an experiment to look at the performance and conduct the analysis of data from the experiment. The design of the experiment was not part of the requirement, instead, it was prepared and data from the experiment under different conditions was to be analyzed. An outcome of the experiment was a professional report from the students after the data analysis. A pre-lab was required and this took the form of a video and a set of multiple-choice questions. After successfully completing that the students could undertake the laboratory which had two sets of data.

The advantage of using video for repeated information was used in a required pre-lab work in a chemical engineering course [65]. The pre-lab was used for not only background information but also for establishing the procedure with equipment as well as important safety factors. The author of the study reported using video to explain the equipment so that there was not the time pressure on the instructor to repeat the preliminary information on the equipment to groups of students, as well as giving the information in the potentially distracting environment of the laboratory. Besides saving time, the video recording provided a consistency to the information given to each student, as well as allowing students to replay the information to clarify any details. There is a benefit too for any new instructor to the course. Collected student feedback indicates the ability to replay and re-watch the information was viewed as a great benefit to using the videos.

A similar use of videos in a pre-lab setting is their use in preliminary training with CAD tools that were available to the students for the course [181]. In this work the author explains the use of computer screen recording to video to help third-year mechanical engineering students get underway in a computer aided engineering course. Alternative approaches used were in-class demonstrations, computer lab training, tutorials in books, CD/DVDs and slide decks. From a student survey 54% of students indicated they preferred to have in-class learning rather than using the videos and 44% indicated they watched less than five of the thirteen videos and 44% indicated they did not watch many of the videos more than once. The transition to available videos allowed comparison in the results between consecutive offerings of the course, without and then with the videos. The result of the comparison of student performance showed "the performance of the students in the course in which the recorded lectures was used was not as good as the students in the course when the lectures were presented in class" [181]. The author does acknowledge that the reasons for the weaker performance could be due to factors other than the availability and use of the videos. The results of this study shows a less than clear benefit to the use of videos. There is the ability to save time, but the availability of the videos does not mean they have been taken on board. One factor to consider though is that the CAD video work was done in 2004 and this was just before the release and availability of YouTube™. Since the release of that Internet platform there has been an explosive use of it as both an entertainment tool and as an educational tool. At the time of writing the use of YouTube™ is extensive, from individuals to companies and news reporting services. A generation of current students will have grown up using YouTube™, including for education. Perhaps the reticence of the students in the CAD study may not be there as strong today.

Another approach to video recording experiments is for this to be done by students, instead of the instructors. The recordings could show work done during and related to the experiment and could be used as a form of video report. In the past such video recording would have required expensive and large cameras and editing could be a challenge. However, the commonly available smartphone that most students carry, or tablet computers can provide easy video recording. Editing can be done on relatively easily available software such as iMovie® on computers, tablets and smartphones. Now, there is a way to record the sight and sounds related to an experiment and to combine footage to illustrate observations, analysis and present findings.

One study of the use is in the laboratories of second and third year biochemistry courses at Monash University [246]. The laboratory component was 3 hours per week in a laboratory and 2–3 hours a week in lectures. Students were introduced to the idea of the use of videos and the assessment approach by a video. The duration of the report was expected to be 5–10 minutes and was to be uploaded to the learning management system. Guidance on the process for creating the videos and including reflective practice was included. One or two weeks were given to video editing. The authors noted the enthusiasm for creating the videos and a perceived increased engagement with the labs. The submitted videos were noted to be of high quality and the authors reported evidence of high levels of critical thinking. Grades were higher for the cohort that submitted video reports rather than a section that submitted traditional written reports. In a survey 59% of students agreed or strongly agreed that the video report allowed them to think creatively. 27% disagreed or strongly disagreed. When asked if the video report was a valuable learning experience 42% strongly agreed or agreed, whilst 27% disagreed or strongly disagreed. 31% were unsure [246]. A key concern of the students that was noted was the time needed to edit and produce the video. This may be a weakness in using video. The time, and potentially the assistance, needed to create the videos has to be balanced for the students. This is a new communication approach and for many, it will take time to learn. The perceived relevance to professional work, over the popular form of report writing, may also be a challenge. So, any educator deploying video reports may need to carefully explain to students the reasons why video creation is included in the laboratory.

Having students make videos has been reported in a materials test lab [180]. Here a GoPro HERO® camera was used as a robust camera systems that could be potentially expendable if damaged during nearby impact, compression, tensile and bending testing. The higher-than-normal frame speed of 120 fps could also be used to help slow down observed changes in the test pieces in the testing apparatus. Students were expected to create videos of their experiments and upload them to YouTube™. An assessment rubric was created in consultation with the students and grading was done in teams. The videos provided a different way to communicate the experimental findings and the speed of the cameras was sufficient to show material changes under time when in the test rigs. Student feedback indicated satisfaction, 73%, with the experiments and 55% believed their technical communication skills had improved [180]. There was raised concerns about the relevance of creating this type of report as it is unlikely to be required to do a video report in future courses, or in employment.

Before we move on from this topic it is perhaps worth considering the use of video-recorded laboratories and the impact on learning using the Kolb learning cycle. A concern can be that the concrete experience is diminished because of the use of the video. The student is not interacting directly with the experiment. They are not using the equipment, not setting it up, nor adjusting measurement instruments and not in the actual presence of the experiment. The experience is all done through the selected choice of the director of the video and the subsequent edit. So how does this impact the cycle? The concrete experience is modified and limited but focused on particular elements that the educator wants to highlight. The video experience may be something that cannot be safely experienced by students, for example the military gas turbine engines. So the video does provide some experience and most likely plenty for a meaningful learning example. The real benefit will come in the reflective observation and the abstract conceptualization phases of the Kolb cycle. Because the video is focused on a particular aspect that the director of the video wants to illustrate, a carefully filmed, structured and edited video will direct the attention of students to specific learning points and the focusing of attention could enhance the reflective observation. Points can be carefully illustrated and attention can be drawn. The view of someone standing a close, but safe, distance may not be as good as a zoomed in camera lens to a specific part of the gas turbine. Students can observe more from the focused image. As well there is the ability to rewind and rewatch to see details perhaps first missed. Collected data from the videoed experiment can be processed in a similar way to normal laboratories so the processing part can be similar. The reflection on the collected details can then potentially lead to improved conceptualization from the edited video.

What about the student-created videos? The need to collect video footage of the experiment can lead to careful decision-making regarding what to film and to not edit out of the report. This reflection on the important elements can lead to a better experience and a more tightly linked reflection. Repeats of the parts or of the whole experiment may lead to a full passage around the Kolb cycle. The reflection may enhance the abstract conceptualization and the active experimentation. The only downside, as mentioned, is the potential time it takes to film and create the video. Here the time and effort spent with processing the video and audio may result in less time on the experiment or other experiments. There could also be a conflict in the purpose, with the risk students focus on the quality of the video production and less on the content.

We can see that the use of video laboratories is a balance. It may be valuable to have the recorded set of data due to safety, uniqueness or difficulty running an experiment. The limited experience can be compensated by enhancement of other areas and it is this enhancement that lies the potential with using this type of laboratories.

3.6.3 Simulation laboratories

A simulation laboratory is another form of the virtual laboratory. Here most likely the simulation will be done on a computer. Computer simulation in engineering has developed significantly over a number of decades, from bespoke simulations using numerical methods and languages like Fortran to sophisticated commercial simulators that can be used in specific areas of design, including computational fluid mechanics, electromagnetic field simulation and electric circuit simulation. The use of such simulation tools in industry and research in engineering also means that students should be familiar with their principles and operation. Therefore, including computer simulation in laboratories and the curriculum is necessary.

The approach to simulations can be split into the student creating the simulation tool, which could be part of a broader awareness of numerical methods and computer coding [38], or using an available tool. A few decades ago the majority of simulation work would be self-created but today there is a growing emphasis on use of more of the industry standard tools. The inherent challenge of any simulation tools is using them wisely within their accuracy and capabilities. As any simulation designer will be aware that using parameters that can cause the simulation to deviate away from acceptable physical reality due to a breakdown in the numerical accuracy or incorrect parameters can lead to erroneous results and jeopardize a design. Awareness of this is an important lesson for students.

Simulations on their own can be a focus of a laboratory course or they could be supplemental to measurements. When combined with measurement the simulation can be used prior to the actual construction and measurement, for example to help select component values in an electronic filter design. Alternatively, they can be used to try and understand results, such as running a computation fluid dynamics simulation of a component in a wind tunnel. If we adopt a Kolb learning cycle model to the laboratory learning the simulation can help with abstract conceptualization, to see if measurements are matching theory, and active experimentation, where simulation experiments can be designed to examine different conditions.

The disadvantages of simulations can include it not being a full substitute to a real experiment, as well as there can be a lack of familiarity with real equipment [22]. The operational advantages are its use outside of a laboratory room, the ability to rapidly change parameters and re-run, safe operation in conditions that could be unsafe in a real experiment and also the cost, which can be less than multiple equipment items for a real experiment. The use of simulations to help with data analysis, visualizing results and understanding the effect of parameters and test conditions on the subject being simulated are valuable.

As already mentioned the array of simulation software has developed significantly over time and the options available to an instructor and their students is significant. In some way the levels of simulation available can follow the pattern of the historical development of software. From low-level coding, through more advanced technical computing tools such as Matlab®, Simulink®, Maple™ and Mathematica, through to more specific software tools that allow specific technical situations to be examined. An instructor three or four decades ago would have students learning to develop their own code for simulations, written in a language like Fortran of C (which is still an option today for learning simulation fundamentals and numerical methods). However, there is an opportunity today to use higher-level tools which is more likely to be encountered in an industrial or research setting. An example study of the use of a chemical process simulator was reported [153] and detailed its implementation in three different universities at different levels, from an introductory university level, through upper year undergraduate, to postgraduate levels. Different simulations were undertaken at the different levels. The feedback from student at the early level found the simulation to help with understanding theory and it was to be interesting. This finding shows that the experience of using the simulator encourages the relating of their observations to theory. This follows a pattern around the Kolb cycle: concrete experience (running the simulation), reflective observation (obtaining and interpreting results) to abstract conceptualization (relating findings to theory). This indicates that even at an introductory level the use of sophisticated simulators by students can be a benefit to their learning, however guidance and instruction on the running of the simulation will likely be needed, which was a finding from the study [153]. With the postgraduate students the simulation exercise examined by the study showed that the students could explore the safety analysis of the chemical process being examined. Finally, the upper level undergraduate and masters level course reported that students gained insight into the control system in the process but were reaching towards the limits of the model's capabilities.

At the early levels of an undergraduate program a course in numerical methods and simulation will usually occur, following appropriate introductory computing and mathematics courses. With the progression in the undergraduate program, students can have access to the higher-level tools and specific modules for specific cases. This will provide undergraduates with knowledge of designing simulations, running them and interpreting the results. So developing knowledge and experience that will be useful and valuable for the students in entering industry or moving onto higher degrees. At the postgraduate level in engineering students will move deeper into their technical area of interest and there more specific simulation work will be done, most likely involving the high level tools in their specific area. A learning pathway for students' simulation knowledge and skill was developed by Magana who approached this by examining the requirements by stakeholders [166]. In the study experts in computational simulation from a range of engineering disciplines and science were solicited for their views, 19 were from industry and 18 were academics. All agreed that there was a need for simulation and computational skills in their discipline. The results of the examination of the views and opinions showed a collection of views on how models can be: i) constructed, ii) used, iii) compared and evaluated, as well as iv) revised at first year (stage 1), mid to final year undergraduate (stage 2) and finally post graduate (stage 3) levels. This could then lead to a derived learning path for engineering students on computational simulation that build in each of the categories (i-iv) at each level. The path is comprehensive and if we just look at the first two stages (the undergraduate levels) of the *use* category (ii) we can start to see the learning progression that had been determined.

For the 'use' category.

> Stage 1: "Students use existing computational models or simulations to comprehend, characterize, and draw conclusions from visual representations of data by evaluating appropriate boundary conditions, noticing patterns, identifying relationships, assessing situations, and so forth."
>
> Stage 2: "Students use simulations at different scales to deploy the correct solution method, inputs, and other parameters to explore theories and identify relationships between modeled phenomena Students use computational models or simulations to design, modify, or optimize materials, processes, products, or systems Students use computational models or simulations to design experiments to test theories, prototypes, products, materials, and so forth Students use computational models or simulations to infer and predict physical phenomena or the behaviors of engineered systems." [166]

The other categories have similar progressions. Presented here is just a segment of the learning progression. Anyone involved in program designs, simulation course design, or having an element of simulation in a broader laboratory experience should consider the learning progression in [166].

While considering the development progression of simulation work it is worth drawing attention to the ideas of J. M. Wing regarding 'computational thinking' [283, 284]. The concept of computational thinking relates to the idea of relating how to do things using computers and ideas from computer science. The characteristics of computational thinking were given as [284]:

- Conceptualizing, not programming;
- Fundamental, not rote skill;
- A way that humans, not computers, think;
- Complements and combines mathematical and engineering thinking;
- Ideas, not artifacts;
- For everyone, everywhere.

We can see then that this goes beyond thinking about computing as just coding, but it has at its heart learning to approach problem solving and tasks with the use of computers and computer systems operate. This idea of computational thinking has entered into schools and has found strong acceptance such that there are books dedicated to helping with computational thinking [73].

If we return to the stages of simulation work that Magana [166] defines in the learning path described earlier, we can see the progress from early simulation fundamentals and coding, towards more sophisticated approaches which may involve design and optimizing through simulation, for example. This is moving students through an experiential learning cycle over both shorter and longer timescales. Within a term's course simulations may build a student's knowledge in different aspects, such as creating and running the simulation, obtaining, visualizing and reporting the results, considering changes to the model to perhaps optimize and so on. However, over the longer term a program needs to have some planning to how simulation experience is developed over different terms and the years of the program. Although there will be a natural development of increasing challenges, given the importance in engineering of simulation work, it would be worth undergraduate program leaders to look at key elements and stages of the simulation experience that students undertake across their programs and ensure there is a range of experience, such as in design, visualization and limits of accuracy for example. This then creates a more comprehensive education rather than the simulation being focused solely on it complementing theory and illustrating principles for the current course. In some way this is taking the holistic idea of computational thinking and applying it to the area of simulation.

3.7 Linking to Situative Learning Framework

Before finishing this chapter it may be worth considering how laboratories fit with the ideas around situative learning. The laboratory environment can be a rich space for learning. Students work within the laboratory room undertaking experiments and related activities. The actions are associated with building, constructing, measuring and testing. They will interact directly and indirectly with student, teaching assistants, instructors and technicians. They may be working with one or more partners to complete the laboratory task. The learning may have started before the laboratory with a pre-lab exercise and after the lab experiment there may be more data analysis and report writing to do. The communication of the findings is often an important part of the laboratory exercise.

When undertaking the laboratory there is the use of equipment, components and other test equipment. Data may need to be found from data sheets, tables of data and other reference sources. Syntax in computer programs may need to be checked if it is a computer-based laboratory. Diagrams are often used in laboratories, whether for construction, analysis or reporting. Analysis of the data may need data manipulation and plotting of results. There may need to be the use of theory and simulation tools to understand findings and results. The mediation process in a lab is rich with different tools and processes.

As the student works in the laboratory and gains familiarity with equipment operation, successful completion of experiments and the processing and reporting of data and findings, there can be a stronger sense of undertaking engineering work and correspondingly the development of identity of being a student on the path to become an engineer. This can be further enhanced with the interaction with supervising personnel, such as instructors, teaching assistants and technicians.

If the laboratory is virtual and perhaps done remotely at the student's residence, there may be an impact on the perspective on the authenticity of the work. To avoid this there may need to be careful design of the laboratory exercise and its objectives.

As the summary of some of the key areas of impact on the situative framework elements of action and interaction, mediation and identity are given in Table 3.1. The richness of the experience of working in a laboratory environment can lead to not just the development of knowledge and familiarity with using technical equipment, but the process of 'doing' engineering and 'working' as an engineer would help develop the skills and identity of being an engineer.

TABLE 3.1
Some key elements of laboratory work related to a situative learning framework.

Interaction and Action	Mediation	Identity
Performing the laboratory exercise	Using equipment and components	Ability to use equipment effectively
Data analysis and reporting findings	Schematics and diagram interpretation. Analysis techniques and tools	Able to interpret and understand technical information
Interacting with student partners, teaching assistants, instructors and technicians	Communication skills (written and oral)	Working with other professionals

3.8 Chapter Summary

The laboratory experience is the classic form of experiential learning in engineering schools and will continue to be an important area for developing experience in equipment use, data collection and reporting of findings. There has been a long-standing criticism of 'cookbook' labs, with recipe-like procedures, and more lab experiences are breaking away from this. We have described some approaches here, including the use of the Kolb learning cycle to provide structure to more 'open' laboratories. Besides the development of lab approaches there has been a rise of interest in a more broadly accessible fabrication-centred facility or makerspaces.

The development of laboratories has also occurred with the growth of computing power and in recent years there has been a growth in the use of online laboratories and simulations, including using professional-level simulation software. These show an opportunity to broaden the laboratory experience as well as assist in providing experience in simulation tools that may be used in the student's future engineering work when they enter the profession. It is expected then that in the years following the COVID-19 pandemic period there will be a number of significant publications on methods for running student laboratories and experiments online.

Finally, it should be noted that as laboratory experiences develop there could be the use of ideas from project work, and the division between the two could become blurred. Projects will be discussed in Chapter 6. If a laboratory adopted more of a mini-project approach, this may be approaching what the Grinter Report highlighted a desire for, of a smaller number of experiments with a focus on analysis and reporting skills (see Section 3.2).

Next, we will look at the main way that students acquire knowledge, in the classroom. Although not usually associated with an environment for experiential learning, classrooms can provide opportunities to include active and experiential learning.

4

In-class Experiential Learning

4.1 Introduction

Experiential learning within a class may seem an unlikely place to find experiential learning. Lecture theatres with a faculty member passing on information through a chalkboard or a projector screen has been a common path to educating students for many years and will likely continue. The common issue with this approach is the risk of passivity of the student, who becomes an audience, with the instructor being the focus of the class. So, can a campus classroom be a place where experiential learning can occur? A conservative view of experiential learning may root this type of learning in co-op placements, project work, laboratories and similar. These are covered in other chapters in this book, but as mentioned in the introduction I will take a broad interpretation of experiential learning in this book.

For experiential learning in the classroom the challenge then becomes how to use a space that may have been designed for the students to 'view' the instructor at the designated front of the room, where the chalkboard or screen becomes the focal point of the room, and the area where the instructor has to occupy. Universities are moving away from rooms like this and we will discuss this in this chapter, as changing the layout of a room can encourage or inhibit alternative learning from the traditional lecture.

In the 20th century lectures may have been one of the most effective ways of transmitting large amounts of detailed knowledge from an expert to an audience. In-class course notes and textbooks were the common form of acquiring course content, especially before the Internet took hold. Today with wireless technology it is possible for information to be accessed off the Internet from inside a classroom during a lecture, making it a tool and a resource. Alternative approaches to learning and delivery can be used. Material can be searched and guests from off-campus can be brought into the class through video-conferencing. It is possible to have students in different locations and even countries work together. Classes can be flipped, such that content can be provided prior to class and material can be discussed and examples given in the in-class sessions.

DOI: 10.1201/9781003007159-4

In this chapter we will look at methods of experiential learning including flipped classrooms and peer instruction. Work in recent years has been looking at re-imagining the classroom structure and this has lead to the idea of active learning classrooms. Since the physical form of the classroom can affect the teaching and learning approaches, we will briefly touch upon this too. To help with the understanding of how the classroom-based experiences can be viewed as experiential learning we will make use of the Kolb learning cycle.

4.2 In-class Experiential Learning

It could be argued that the classroom is commonly a less than optimum opportunity for learning, especially learning by experience. Often a large number of students come together with an instructor to form a gathering of people that can focus on a particular topic and share views and insight guided by the instructor. Unfortunately, this opportunity is often lost (and I am as guilty of this as any instructor) and instead of sharing of thoughts and ideas there is a move to having the instructor take control, be the majority source of information in the room (there may be student questions or comments) and students take a role of an observer and collector of the transmitted information. Notes may be taken and handouts may be annotated. If we consider the Kolb learning cycle we can see that the direct experience of the student is limited and the knowledge passed over is most likely fixing the learning in the abstract conceptualization stage. Some may argue that the Kolb cycle is invalid here, and should not be applied, since there is little concrete experience. Through direct transmission for the instructor the student is acquiring knowledge, which is likely unfamiliar to them and most likely for engineering, in an abstract form using mathematics, diagrams and physical concepts. Demonstrations may be given, which will help, but the experience will not be as strong as if the student was more actively involved. In this section we will look at some approaches that have been reported that break from the common lecture style and that involve more active learning and so provide an experience that can help the students start on the learning cycle. This is a efficient way of conveying the conceptual information and it is different to learning solely from a book as students can ask for clarification when needed.

One way to break from the tradition is to consider the use of case studies. These have been popular in business schools and can translate well to engineering where in practice an engineer will find themselves with a problem that needs solving. The application of case studies in engineering has been used by Raju and Sankar [217]. The authors worked with members of a steam power plant where at one time a turbine had had a vibration issue that caused it to be stopped. Engineers at the site had differing opinions on the cause of the vibration. One engineer considered there was an imbalance and repair was a little under $1M whereas another, whose role was predictive maintenance considered the problem was different and recommended an immediate restart. If this second option led to damage to the turbine a replacement could cost $19M. Using information from the plant supervisors a case study was developed. Students were split into groups and were presented with the work. Two groups took the different roles of the engineers recommending the two different solutions (repair or restart). One group took the role of the plant manager that had to make the decision on the route to the solution and another group looked at alternative technologies. The groups were then faced with a range of problems from financial, technical and a management perspective. Students could use the provided data, undertake research and get first hand information from a visiting plant engineer involved with the case. The case was finally debated in class. Student feedback on the case study was collected and this was found to be liked and viewed as useful and challenging.

If we consider this case-study work approach we can try and see how the cases study fits with the Kolb learning cycle. Although the students did not directly experience the turbine's problems or the plant manager's situation, the case study was detailed and authentic making it a well-presented problem that posed not just technical challenges but it included financial implications too. Reflection would come with examining and considering the information presented and researched. This was further complicated by having the groups take on a role and so defend a proposed solution, which could bias the way the evidence was considered. Conceptualizing would occur with the information analysis and this would also consider the financial and plant management implications. Risking a restart without alterations could have severe financial implications along with a period of time when the turbine was inoperative. The proposed solutions and the degree of argument for the defence of the role's stance would come together with the active experiment stage when proposals were evaluated and decided upon. The papers states some challenges were made on the technical details so here was a situation where students were evaluating and challenging technical assessments by other students.

This reported case study indicates the different way the class time could be used. Research and preparation were required. The factors to consider went beyond solely being technical. The situation was not abstract or devoid of context either. The students had an opportunity to meet an engineer involved. The authenticity and detail could draw the students in and their role was far from passive in the learning exercise.

4.3 Flipped Classrooms

One recent change in classroom teaching that is gaining popularity, is the idea of the flipped classroom [199]. Instead of using scheduled in-class time for transferring information and outside class for problem analysis, the order is switched. Information is studied before a class, so that in-class is where problems, questions, discussions and clarifications can be worked through. In many ways this type of flipping the class has been facilitated by technology. Learning management systems have allowed resources to be structured and shared, as well as lectures or experiments can be video recorded and made available to students for viewing prior to the class. This is not to say that technology is necessary, as I suspect many innovative instructors will have used guided reading from textbooks and papers to facilitate in-class activities that use the material learned in the readings. So the technology is a tool, like a book, to facilitate the learning, prior to class. This moves the 'first time' learning to more of a solo and individual setting, and then the use in analysing or applying the knowledge moves to the group setting. This is the reverse to a traditional model where the class is often a lecture providing information to an assembly of students and problem questions are done outside of class, often by student on their own.

In the flipped approach the in class session can then be more varied. Students can raise any difficulties they have with understanding the material and the instructor can help clarify. Examples could be worked through and discussed. As mentioned in the previous section a case study could be explored. Options open up with the freeing of scheduled class time moving away from instruction.

If we consider Bloom's Taxonomy (or a revision) we can see that for a flipped class the initial levels, like remembering and understanding, can start outside of the class and further application and analysis of the material can occur inside the class in an assembly of students all working on understanding the material. This provides opportunities to move to higher levels in the taxonomy in a collective. Being together to hear the questions from other students and answers from the instructor allows perhaps a better staging of the material. Alternatively, we could consider the Kolb learning cycle. By having prior examination of the course material before the class, the knowledge is in place for the abstract conceptualization stage. So any experiential activities in the class can be reflected upon and then related to the conceptual understanding of the material. In short the prior learning allows a boost through the learning stages (the activities act as a form of quicker feedback, compared to mid-term or end-of-term exams), whether one considers the Bloom or Kolb models.

The use of video or similar (such as an audio recording) allows a lecture to be captured relatively easily for viewing and students have the opportunity to view and review the information in the lecture. There is an opportunity then to use the instructor's and students time to develop the material and question it in the classroom, maximizing the scheduled time. This does require time for the instructor to record the lecture and for the student to view the material.

Studies have been undertaken to determine if there are any benefits to the flipped classroom approach. These studies can be broken down into reports of the implementation of flipped classrooms such as [274] and [53], or larger survey or meta-analyses of a number of individual reports, such as [147] who looked at 24 studies from 2009 to 2014 on flipped engineering courses, [163] who considered 29 comparative studies between flipped and conventional classes, from between 2008 and 2017, and [145] that looked at 62 flipped classroom studies between 2000 and 20015. These examinations showed that in general there were benefits in performance and student achievement. Kerr identified that there was favourable reception to the flipped classroom approach including liking the use of in class time for problem solving, though there still seemed to be some preference to having face-to-face lectures [147]. Lo and Foon's study carried out significant statistical examination on published data and they reported quantitative improvements in students achievements with flipped classroom approach [163]. Karabulut-Ilgu et al also showed that there was a rise in publication, linked to a rising interest, on flipped classrooms around 2012 [145]. They also noted the need for more research in the changing of learning as well as making a call for more of an examination of the move from the benefit to learning achievement but to also the development of professional skills.

This detected improvement in these survey or meta-studies, which indicate a general performance improvement, then raises the question what aspects of the flipped approach provides the benefits? Lo and Foon examined this and determined that the highest impact they observed was with an in-class format of starting with a brief review, then some individual tasks and also small group activities [163]. From these stages we can see the review could provide a reinforcement of the concepts of the material (valuable for abstract conceptualization); the individual tasks could provide concrete experience in problem solving or a case study, drawing in reflective observation followed by abstract conceptualization. The small group activities bring in more experiences as well as the other parts of the Kolb cycle. Discussions could instigate reflection, and into relating the experiences to the underlying concepts (the abstract conceptualization phase). This can cause re-examination of the problems by the individuals and group, leading to active experimentation with the problem.

As mentioned, if we use Bloom's model, we can see the remembering and understanding model from the pre-class preparation is now extended with the in-class activities that move up the taxonomy into applying, analysing and potentially beyond. In a simplistic reasoning it can be considered that the pre-class work and the in-class problems provide a closely following sequence of events that is different to a traditional class, where material is collected in a lecture and potentially not reviewed until some form of coursework or assessment. The problem being the assessment may be considered later in time. This differs compared to the flipped sequence of events, where preparation, onto review, through to individual in-class tasks could be quicker. This shorter cycle between material and problems may provide better depth of learning than the potentially longer period of time between lecture and assessment.

Questions can be raised about the form of the flipping stage and of the in-class stage. What are the activities within each that can maximize the benefit to students? For the pre-class stage the ability to self-pace is noted as being an advantage [163]. There needs to be consideration of different types of learners [102], such as those that need to see the whole picture – global learners, or those with dyslexia. The use of video for pre-class work can help with accessibility and allow the video to be played slower or repeated for those who first language is not the same as the language of the course delivery. Videos are recommended, but should not be too long, not a full traditional lecture length. Instead, it could be reduced to a shorter length with just the key (uninterrupted) information and then potentially further subdivided into shorter segments for easier access and viewing, such as 3–5 minutes [102]. In a study of basic science for medical students it was found that an optimum length was 10–20 minutes [42]. The authors also point out that the issue of very short videos results in many videos being produced which leads to an organizational problem for both students and course organizers.

The approach to video production is something that could be varied. From videoed lectures to mini-lectures and demonstrations, worked examples and guest speakers. The variety could be rich and this may need some level of organization and arrangement into a cohesive route through the course. Guidance on how to approach this more open and self-paced approach to learning may be needed. The in-class sessions should be one route for the student to get feedback on how effective their pre-class preparation is, but additional feedback and guidance may be needed and sought.

Within the in-class session we have already noted the sequence of brief review, individual tasks and small group activities. This provides a guideline, but each instructor will have their own specific effective methods for reinforcing the learning. Furse and Ziegenfuss give comprehensive guidelines for flipping classrooms, with their attractively titled paper "A busy professor's guide to sanely flipping your classroom: bringing active learning to your teaching practice" [102]. In the next section we will talk about one successful in-class technique called peer instruction developed by Prof Eric Mazur [176]. The freeing of the in-class time from the transfer of course content is liberating to the instructor, who now has the ability to use the time to pay attention to knowledge development and to move the students higher up the tiered Bloom's taxonomy towards the analyze, evaluate and create levels. The instructor can also play to their own personal strengths. It is worth taking a quote from Ken Bain's book 'What the best college teachers do' [21]. In the epilogue he says

> "Perhaps the second biggest obstacle is the simplistic notion that good teaching is just a matter of technique. People who entertain that idea may have expected this book ['What the best college teachers do'] to provide them with a few easy tricks that they could apply in their own classrooms. Such ideas make enormous sense if you have a transmission model, but it makes no sense if you conceive of teaching as

creating good learning environments. The best teaching is often both an intellectual creation and a performing art."

As the flipped classroom is moving away from just transmitting information the focus on the creativity in the classroom will be increased. The Flipped Learning Network provides 'four pillars' to guide educators with flipped classes [198]. They follow the acronym FLIP. In order the pillars are: **F**lexible environment, **L**earning culture, **I**ntentional content and **P**rofessional educator. The flexible environment relates to changes that can be made in the class to help facilitate the learning, for example adjusting to help small group work. There is a reason why active learning classrooms can look quite different to the traditional lecture theatre. The learning culture pillar relates to the loss of control of the learning by the instructor, who is no longer the lecturer, but instead the emphasis is on the student to engage in a deeper way with the course. The intentional pillar acknowledges that instructors need to guide the learning and develop materials carefully and with intention. The final pillar acknowledges the professional role of the educator and the need to reflect and develop teaching abilities that help develop the flipped learning environment, where they are less likely to be at the centre of attention during learning (unlike a lecture), but instead will be observers and guides in the process.

The previously mentioned paper by Furse and Ziegenfuss [102], can be a good starting point for anyone wanting to move a class to a flipped form. There are practical details and advice given, including move over to a flipped form in stages. One example they suggest works around splitting a term into three parts and starting in the first term with active learning approaches in class. In the second term video examples could be provided and in the final third video lectures can be provided [102]. If videos are used, as mentioned earlier, then there needs to be careful organization of the library of videos. Learning management systems can help to organize the pre-class materials of whatever kind. Furse and Ziegnefuss also encourage the use of soliciting for feedback regularly on the 'muddiest point', being the points in a 'lecture' that they find the most difficult [102]. Soliciting for regular feedback can help students from feeling that they are learning in an isolated way on their own. By using a series of feedback questions there can be encouragement to have the students reflect on their learning through the course [102] and as well the instructor gets valuable feedback for improving the classes in future. The feedback questions can the students feel supported and as well it can help with their own reflection and articulation of how the class is going for them.

Despite the overall evidence from meta-studies that there are measurable improvements in student achievement [163], some who have studied flipped classrooms report there was not a discernible increase [3, 243]. However, the use of grades is perhaps not fully quantifying the benefit as Smallhorn reports. Despite a study that shows not significant change in the learning outcomes of students, the study by Smallhorn did report a perceived increase in the engagement level (measured through attendance and assignment submissions), which can translate to increased retention in the program or course [243].

Are there potential disadvantages to flipping? The materials need to be well organized and succinct. This will help guide the students through the material in an efficient manner. A failure in this could lead to confusion and the risk that students spend too long trying to progress through the material. Multiple flipped courses in a term of study could move a significant amount of work in unscheduled timetable time, with classes still taking up the formal timetable of events. Overloading can become a risk. There is a similar disadvantage that the instructor will have more work to do, as videos or similar materials for the unscheduled timetable will need to be produced, as noted by Lo and Foon [163]. However, done correctly there are noted advantages of flipping. The self-pacing through the instructor-provided materials, helps students. When the class comes together the activities undertaken can build on the pre-class learning and help clarify any difficulties in understanding. As mentioned, these activities allow the instructor to guide students to progress up the levels in a Bloom's taxonomy model, or provide more experiences in a Kolb learning cycle. The challenge to the educator can be with both the pre-class material and including effective methods in the class for learning. One method that could potentially be used in peer instruction which will be the topic of the next section.

Before moving on it is worth mentioning that the recent interest in flipped classroom approaches are still relatively new (since the about 2010 onwards). This is a developing area and one where more common practices will perhaps emerge. With the COVID-19 lockdowns many universities will have adjusted to online delivery, sometimes with synchronous lectures and others asynchronous, or a blend of the two. This will have encouraged many more educators to try flipped classroom approaches, correspondingly exposing more students to the mode of learning, as well as opening up the possibilities of flipping their classes to educators. The outcome may be an acceleration of the adoption of flipped classrooms in the future. It should also be noted that many of the metadata studies point to the need for more quantitative analyses to see the benefits of flipped classrooms.

4.4 Peer Instruction

Problem solving through peer instruction is centred around students attempting problems that check their understanding and then to facilitate discussion between students to clarify their answers and understanding. It is an elegant method that can not only help student learning but provide rapid feedback to the instructor on misunderstandings and areas that need clarification.

If a class is flipped what can be done to replace the lecture when the class meets? One well-documented and studied approach is peer instruction developed by the physicist and Harvard Professor Eric Mazur [176]. Mazur became concerned when he noticed students in an introductory physics class seemed to have less of a solid understanding of the physical concepts underpinning the calculation problems they were often accomplished at solving. To resolve this he developed a specific flipped approach to his teaching, which was developed many years before the more recent trend and interest towards flipped classrooms. His book *Peer Instruction: A User Manual* [176], describes his path through developing this approach to teaching and learning and carefully details the process and its effectiveness. This book by an expert physicist and educator is a recommended read and here only the basics of the process will be given.

In peer instruction the students in a course are entered into a flipped classroom type situation [176]. Prior to class material is given to the students to study. Mazur used readings. Then in the class he would give a brief review of the material for perhaps 7–10 minutes and then he would supply a ConcepTest for another 5–8 minutes. This would be a question that he could set that was carefully designed to test the conceptual understanding of a principle or aspect of physics. Not a 'plug and chug' problem. The answers would be in a multiple-choice form. Having been presented with the problem students were asked to determine the answer, working in silence on their own. No conferring was allowed. Then there was a poll to see what the choice of answer was and importantly the distribution. After the selection distribution was revealed to the student they were asked (without being given the correct answer) to discuss their solution to the question with a neighbour to convince them of their answer. If the answer was overwhelmingly unanimous and correct this repeat stage could be skipped, but a good ConcepTest problem would result in perhaps two answers being more strongly featured. After the discussion between students (over their approaches and which was correct) had gone on for a designated time there would be a repeat poll. This time the correct answer hopefully become the more favoured choice. If there was a convergence in the correct answer then the instructor can move on to another principle or concept in the course material. If there was indications of confusion or a lack of insight the instructor could return to the material to help clarify the misunderstandings. This two-poll process allows a student to first solve a question on their own and then in a second round allows the student to explain and potentially defend their reasoning for the selection, whilst listening to another student's explanation. Between the peer discussion there is hopefully a clarifying of the understanding, either changing an opinion or reinforcing the answer. Here is where the peer instruction occurs, with student clarifying their understanding or having a misunderstanding exposed. This use of a well-crafted question need only take a few minutes so in each hour of class time can contain about four of these reviews and question-poll modules.

Peer instruction is one of the older and more well-documented approaches to a way of flipping a class. The form has become popular and consequently has been studied. Brooks and Koretsky have examined the discussion process in peer learning in a chemical thermodynamics course and have found that the group discussions help with deeper explanations to the answers to posed questions [44]. They report an increase in information-based reasoning rather than a more heuristic experience-based judgement. When they examined the role in showing the interim answer between the two rounds of questioning, they report seeing no difference in the patterns of change in the answers when the responses were shown or not, but they do note an increase in the confidence level. Smith et al have also examined the discussion process when using a peer instruction approach [244]. They note that student group discussion helps with understanding and determining the correct answer, even when a group is composed of students who had initially selected the incorrect answer. Here is an example where students are learning through a constructivist approach rather than solely through a transmission of knowledge.

The benefits to peer discussion perhaps goes further than just peer instruction and active learning. The broader use of peer discussion may help in all areas of experiential learning. Having the chance to discuss and share thoughts on the experience and learning may be a key element to learning. Whether it is the approach the peer takes to discussing the knowledge, or the acceptance of the peers' opinion and understanding, such that it forces a consideration to the validity of the peer's point of view (which may not occur with ideas and concepts from an instructor), it seems the discussion is an important part in some learning approaches. However, there are reports that instructors should not be naive and thing that all discussions are following in an effective manner to facilitate learning. James and Willoughby [139] recorded student conversations in introductory astronomy courses and revealed that of 361 recorded conversations 136 (about 37%) where what were called 'standard' conversations that an instructor would expect [139]. The remaining, nearly two-thirds fell into three areas according to their categorization, i) 'unanticipated student ideas', such as discussing incorrect ideas unrelated to the question content; ii) 'statistical feedback misrepresenting student understanding', such as guessing or based on incorrect information; and iii) 'conversational pitfalls', including not even starting discussions. The number of identified 'conversation pitfalls' was similar to the number of standard conversations. So, instructors should be clear that a room of discussions may not all be focused in the way they want, but it is perhaps an improvement on a silent audience in a lecture room.

4.5 Classroom Design

Before leaving the area of in-classroom experiential learning it is perhaps worth highlighting the physical layout of classrooms can affect in class discussions [203]. The majority of lectures rooms were designed with a central focus on the instructor and all the student desks are focused towards the front and the instructor's position. This sets up the instructor to be the main communicator in the same way a stage is set up for a theatre performance. If an instructor is wanting to turn the tables, often quite literally, so the students face each other this can become challenging as seating and tables are very often fixed. With the rise in active learning there has been a corresponding attempt to change the formal learning spaces, so that students can more easily interact with each other and share ideas. These spaces are known as active learning classrooms and are becoming more common across university campuses.

Active learning classrooms can take various forms but most commonly they allow students to sit facing each other and often have access to large monitor displays and whiteboards for group work [205, 218]. With the focus and a floor-plan change away from the instructor being the point of attention, there is sometimes a need to have a way for there to be an easy transition from student-to-student focus to the instructor and back again. To do this swivel chairs can be used, so students can turn, and large monitors or projector screens around the room can be switched to display the instructor's presentation. So there can be at least some smooth transition from the peer to peer communication or an instructor-focused session in the class. For larger classroom, traditionally large lecture theatres there is a move towards having tables which are on tiered levels. Students then can sit in groups facing each other, allowing group discussions but the positioning and the tiering allows the conventional stage position to be the focus. Each table could have a microphone that can be switched into the room audio system so ideas from different tables can be shared with all in the room

The advantage to having this re-configuring of the conventional lecture more easily facilitates discussions and active learning. It is no longer the person sitting directly in front or to the side of you that you can have discussion with. The disadvantage becomes the cost for the extra equipment and the change in seating and tables. It often results in the slight reduction in the maximum capacity of the room which sometimes can be an administrative issue.

The aim here is not to digress into the these design and features of the rooms as there are many resources available discussing this (for example [203]). It is recommended for faculty to search out and possibly request these rooms for a class, if they aim to feature more active and experiential learning. As well, if the opportunity arises an instructor who successfully uses flipped classroom techniques or similar can be involved in redesigns of rooms, or the design on new learning spaces. As architects and designers are requested to offer alternatives from the focused lecture theatre and provide options towards more flexible spaces that can facilitate a variety of learning approaches, then dialogue between the learning space designers and instructors will be needed.

4.6 Case Study

Activity: Engagement through open-ended questions.
University: University of Auckland, Auckland, New Zealand

Peter Bier is not only an accomplished unicyclist and juggler but as a holder of New Zealand's National Tertiary Teaching Excellence Award (2109) he is a recognized outstanding educator. He teaches mathematics at the University of Auckland, Dept. of Engineering Science. He has developed and implemented an approach to engaging in mathematics that has been used at both high school and university undergraduate levels. The approach is to pose to a relatively open non-trivial question (generally with little or no numerical data in the question) to a group of students and to ask them to provide answers in report form within a day (usually seven hours). Questions used include

> "How far do you have to dive underwater to be safe from gunfire?" and "If a severe Tsunami warning was issued, how long would it take to evacuate the 13,000 people who live on Te Atatu Peninsula?" [30].

This project is done as part of a first mathematical modelling course and the project is done in groups of about 4 students. The questions are rooted in real world problems and pose specific questions that may not have easily found solutions through research. Instead the students will need to determine approaches to solving the problem, research and gather data and then explore possible solutions. When they feel they have their best solution within the time they have to write a report presenting their solution. Besides the undergraduate modelling course this type of one day problem solving exercise has been used as a national high school mathematics competition.

The challenge of the problem allows for a variety of approaches to be taken and the modelling process to be repeated over the duration of the exercise. The requirement of the report allows for the documentation of that report. The time limit, which includes the report writing time provides students also with a time management challenge. Bier managed the ability of the teams by dividing the class into four based on their ability in a diagnostic test. Each team was made up of randomly selected individuals from each of the quartile levels. He has also tried to manage the gender balance in teams by trying to avoid having a single female in a group, and having at least two or none. To ensure degree of commitment and working together on the problem, the students are told they will all receive the same assessment mark. The students also provided peer and self assessment to the instructor.

A study of the student perception of the course was conducted through a series of surveys given: i) just prior to the project, ii) just after and iii) again towards the end of the course when all assessed work had been submitted [31]. The survey generally showed after a high opinion on the exercise helping understanding prior to the project day, it would fall somewhat directly after the project and later would tend to rise again, but not to the pre-project level. Peer assessments indicated that there was some belief that some students put in more work than others. Feedback was also given that the vagueness of the question was challenging and some students wanted a more well-defined problem. This however would go against the instructor's aim of the exercise which was to look at how appropriate assumptions could be made in modelling.

4.7 Chapter Summary

In this chapter we have taken a broad view on experiential learning. Much of what is written here can be considered to be 'active' learning, but it can be argued that an appropriate and well-designed in-class exercise can be a valuable piece of experiential learning. Using scheduled class time to analyse a case study problem, or work on a vague and open-ended problem, can let students experience the starting point of a problem analysis for a professional engineer. The described case study from New Zealand provides a good example of implementation and the learning. One of my own earliest forays into such an approach was in an optical physics course, when rather than having students investigate complex optical lens systems I posed the question "why is it blurry if we open our eyes underwater" and I required calculations to show the answer. There often was a sense of achievement when students arrived at an answer through use of the analysis methods, along with the optical structure of the eye, given earlier in the course. I then would follow up with another question, "how is it that some birds and animals could see effectively in air and underwater?" This then encouraged students to postulate how the optical system of an eye could adjust above and below the surface of the water. Without extensive biology knowledge the physics students could come up with ways there could be the adjustment and they were often surprised to see the biological details as to how nature had evolved to facilitate the adjustment [250]. It was the surprise and the sense of empowerment the students had when they could analyse such a fundamental question that made that tutorial question one of my favourites of the course.

Perhaps it does not take too many thought provoking exercises in a class (flipped or not) to encourage learning. Working in groups students can share ideas and ideally dispel misconceptions. From a constructivist learning perspective the models of understanding the student's have and share are being tested and refined. To facilitate the sharing, as well as facilitate group work, then the traditional rowed classrooms that face the front of the class where the 'lecturer' stands, need to change. Many institutions are modifying and building new classrooms to facilitate active learning. The investment in time and infrastructure to move away from the traditional lecture class is significant, but the benefits are worth it.

In the next chapter we will continue with the ideas of changing the conventional way of learning and look at the ideas of problem-based learning, the conceiving-designing-implementing-operating approach to learning, as well as project-based learning.

5

Problem-Based Learning, CDIO and Project-Based Learning

5.1 Introduction

As mentioned in the previous chapter there has been a move away from the traditional lecture format and explorations of different approaches to aid student learning. Some of these approaches have grown in popularity. In this chapter we will discuss some of the more popular approaches in engineering education; problem-based learning, conceiving-designing-implementing-operating or CDIO^{TM}, project orientated learning and challenge-based learning. Each of these areas has a distinctive focus but the common aspect is actively involving the student in undertaking problem solving, design work, projects or similar activities. Having students learning through these methods can help students develop their abilities and experience in ways that will help them in the work place and in a manner that employers are looking for in new hires.

Because of the variations on the schemes, in this chapter we will briefly outline some of the key features, with examples. It will also allow the reader to see the range of approaches. The aim is not to go into any one area in depth, but instead to provide basic outline on the approach and connections to experiential learning. This can help the reader unfamiliar with the various topics to have an overview and to see if any particular approach catches their interest for their own teaching or further research.

5.2 Problem-Based Learning

Perhaps one of the most popular learning schemes in this chapter is problem-based learning or PBL (note care with this acronym as it is sometimes used for project-based learning, and this can lead to debate as mentioned by Beddoes et al [26]). This approach came mostly from the education of medical

DOI: 10.1201/9781003007159-5

professionals and originated from the medical schools of a few universities in the 1960s and 1970s, including McMaster (in Canada), Maastricht (in the Netherlands), Michigan State (in the USA) and Newcastle (in Australia) [24]. The impetus for the development and use of PBL was due to dissatisfaction with the existing medical education approach [24, 131]. Concerns included students being ill-equipping to work in a clinical setting, and their forgetting of fundamental knowledge, as well as a recognized need to develop problem-solving and life-long learning skills [24]. The success of this approach led to other medical schools introducing the approach of PBL, then spreading to other areas, both inside higher education as well as at primary and secondary education.

Pioneering adopters of PBL into engineering education were in mechanical engineering at Imperial College London [57] and chemical engineering at one of the PBL founding institutions, McMaster University. The reasons for the use of PBL in chemical engineering at McMaster were [287]:

- a demand on putting more specialization content into the curriculum,
- the need to develop student's problem-solving, communication and life-long learning skills,
- the need to move from teaching to learning.

The attraction to PBL resided in being able to cover a range of materials, whilst at the same time trying to make students more independent learners and solvers of problems. So better preparing them for professional life. The acknowledged challenges of the approaches are [57, 287]:

- acceptance by faculty (who have to adapt their delivery) and students (who have to undertake a different approach to a course compared to conventional courses),
- adjusting the content breadth, which may be less than a lecture-based course, but this is compensated by the student learning how to better approach self-learning,
- developing appropriate problems and the right number of those problems.

What then constitutes PBL and how is it implemented? For these details we can draw on the material from medical education such as [286].

The form of a particular implementation may have some variation, but the basic form is to have the class split into groups which are reported as varying from 4 students [57] to 10 [286]. Those groups could then have roles assigned including a chair (to organize and guide the group meetings), a scribe (for recording notes) and then the various group members. A further participant is the tutor, who normally will be the course instructor and they may be joined by others depending on the size of the class. The problems have been decided before by faculty and could be presented in a variety of forms. In medicine that could be as a written case, simulated patient, computer simulation or video recording [286]. The variety shown here from medicine, gives an indication on how engineering can provide varied problem cases. Once the students are organized and the case has been presented the self-directed learning can begin. One approach originating from Maastricht is for the students to follow seven specific steps in the process [260].

TABLE 5.1
PBL seven-step approach. Based on [260]. The first five steps make up the preliminary discussion.

Step	**Details**
1 Clarifying Concepts	Clarifying terms and concepts Asking and giving explanations
2 Defining the problem	Proposal for definition Formulating the problem
3 Analysing the problem and brainstorming	Listing relevant aspects Seeking clarity Avoiding exclusion of possible explanations Collecting alternatives
4 Problem analysis. Systematic classification	Linking aspects and explanations Checking clarity
5 Formulating learning objectives	Use knowledge gaps to formulate learning objectives Link with problem analysis
6 Self study	Scheduling. Selecting sources Study sources Check understanding
7 Discussion	Reporting and presenting Discussion of areas lacking clarity Check meeting learning objectives

Table 5.1 shows the seven steps and further elaborates on these points. The first five steps can be considered to be the preliminary discussion. Step six will be when students can work on their own, investigating and collecting information and the final and seventh step is when knowledge can be brought together on the problem and there can be an assessment as to whether the learning objectives have been met. The role of the tutor is to help with any guidance that may be needed – this work is to be student driven and not tutor driven. This can include helping in the guiding of the depth and relevance of the investigation into the subject matter, along with clarifying misconceptions. Along with the gentle guiding, where needed, there will need to be observation of the group and its approach and findings, which can then help with feedback at the end. This process is repeated for a number of problems. Cawley reports

using six problems in his early work of using PBL in engineering courses [57], whilst Woods reported one in a junior course and five in a senior course with each problem lasting a week [287].

The starting problem becomes important at it sets the stage for what the students will focus on and what they will learn. The approach a group takes to defining the problem and identifying the key points for investigation is also important, as it sets the compass to solving the problem and becomes a guide through the learning. The determination of the learning objectives is perhaps where the tutor needs to provide careful monitoring and guidance. Once this preliminary stage is complete then the individual students embark on their investigations for information to solve the problem. Once collected the student has to prepare to present the material, then share and discuss with their group members. This is a distinct contrast to the use of lectures to deliver the material. It can be expected that the duration to gain the information being learned will take longer than if delivered by a lecturer, however, it is the process and retention of the knowledge that can counter-balance a reduction in the amount of content covered. The student, along with a group of peers, learns how to move from encountering a problem, to collecting the knowledge, solving the problem and then communicating the details. The process of establishing what has to be learned and then reporting on the learning brings together skills that can help with life-long learning skills, working on problems, working independently and in a team, assessment of information and its communication. These extra benefits seem to off-set any reduced coverage of the material and may significantly benefit the student in future professional life. To help prime the development of these skills Woods indicated that in their pioneering work in using PBL in engineering courses they provided explicit training prior to the problems being introduced [287].

As it is the starting point for the process, what needs to be considered about the creation of a problem in PBL? This can be dependent on the field and the level of course. A problem in civil engineering may be about a specific structural design, whereas an electrical engineering project may be focused on the performance of a particular circuit. There could be differences in the presentation of the problem, for example a design problem or something related to specific data, or a simulation. The knowledge required will also be a factor. Although PBL is intended to create knowledge, it does require prior knowledge as a base for the development of further knowledge. The more background the students need to understand to tackle the problem, the more of a challenge it will bring. There also has to be consideration of the time it takes for students to work through the PBL problem. Jonassen and Hung point out that there is a domain space for for the problem that has the axes of difficulty and structure [142]. A problem can be located somewhere on a simple to complex axis, as well as being highly structured (perhaps examining a programming code syntax problem) or having little structure (perhaps a open design problem), see Figure 5.1. A created problem can then be located within that domain space. The positioning of a problem in the space, represented in Figure 5.1,

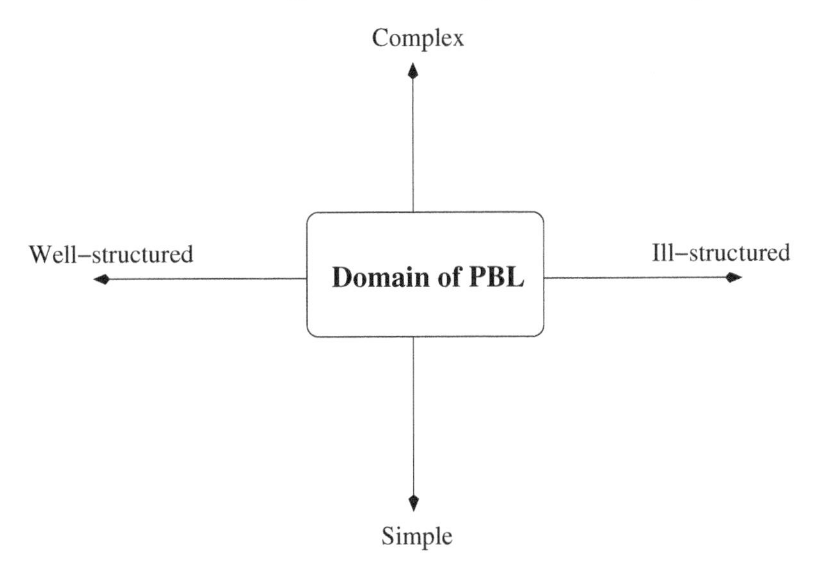

FIGURE 5.1
Domain of PBL after Jonassen and Hung [142].

can set the level of challenge to the students in the PBL exercise, as well as set the type of experience they will have in trying to gain the knowledge to solve the problem. Any PBL problem design will need to consider that space when creating the problem. A sequence of problems may be placed at different points within the domain space.

What then are the characteristics of an effective problem? Table 5.2 is a compilation and adaption of some of the characteristics that Wood [286] and Jonassen & Hung [142] have reported. Both sets of authors indicate that the problem should be open-ended and Jonassen & Hung indicate it should be complex [142]. So this indicates the problem starts to move into the top-right quadrant in Figure 5.1. How far into that quadrant has to be tempered by making the problem appropriate to the course, its level and the prior knowledge of the students. Making the problem authentic is noted and that can link with the idea that any basic science knowledge learning needs to ideally be made in terms of a discipline-based problem. Motivation is another factor. So, problems that encourage participation in knowledge finding are desired, which returns us to the need to be appropriate in the challenge and to it being authentic.

The use of PBL in engineering education has grown from the initial work described by Cawley [57] and Woods [287], although perhaps not as widely adopted as with medical education. A review of the forms of PBL implementations in engineering has been recently published [62]. There has also been books produced on the research of this learning approach [79] and the

TABLE 5.2
Characteristics of an effective problem based on Wood [286] and Jonassen & Hung [142].

Wood	Jonassen & Hung
Learning objectives to be consistent with faculty learning objectives	Open-ended, ill-structured
Appropriate to the stage of the curriculum	Complex including: - challenging and motivating - opportunities to examine from multiple perspectives - adapted to student's prior knowledge and readiness
Scenarios have cues for discussion	Authentic
Problem is open	
Scenarios should promote participation in seeking information	

various international practices [111]. For any reader interested in the details on implementation these books and reviews are valuable resources.

Chen et al do note the diversity of approaches of PBL in engineering and that is a focus for their review [62]. Also included in their study is an examination of some of the challenges that can be encountered that cover the areas of: training for both instructors and students, appropriate assessment methods, lack of time and support for implementation, the challenge in designing a PBL course and cultural barriers such as language or different approaches to learning.

There has been work that has been noted that PBL implementation in engineering education is different to that used in medical education and as well there are limitations [208]. This study notes that a problem-based approach to learning is closer to what a physician does when working professionally than an engineer. Instead, project-based learning or design-based learning are arguably closer to what an engineer encounters in their work. However, PBL can be considered to involve project work and this was included in Chen et al's review of approaches to PBL [62].

If we consider the theoretical aspect or PBL it can be seen that it clearly fits with the constructivist way of learning. The students evaluate the problem, determine what they need to learn, and have to find the knowledge to fill the gaps. The problem is usually ill-defined and they work in groups aggregating and discussing the information and knowledge that they have found. They build their learning and direct themselves over the period of the problem, with the tutor being a guide and facilitator. The sharing of the knowledge with the group brings together other student's findings, perspectives and understanding. This clear link to constructivism has been reported in a number of papers, for example Savery and Duffy [230] and by Kantar with regard to nursing education [144]. Hendry et al did a detailed consideration of constructivism and PBL, using an examination of a specific PBL implementation in a course in medicine at an Australian university [122]. From this they highlight the need to arouse student interest in the learning, reduce anxiety and maximize individual self-efficacy. For PBL they note this can be achieved with realistic problems, support from tutors for reflection and cooperation, sufficient time for the independent study and alignment of the assessment with the learning objectives and implementation of the teaching [122].

The link with a constructivist understanding of PBL is strong, but throughout this book, we make use of a situated learning framework to understand approaches to experiential learning. How then does a situated learning framework fit with PBL? If we consider the elements of action and interaction, mediation and participation plus identity we can start to examine the aspects of PBL down relating it to a situated learning framework. At the heart of PBL is the problem that initiates the learning. Depending on the problem type it will set the type of interaction and participation, for example if it is a simulation, a data-based problem, a design, or laboratory-based problem. The immediate group members and the tutor will be direct participants and form the core of the interactions. There may be information and literature searches that could involve librarians or technical staff, other participants. The problem may require the examination or development of diagrams, mathematics and computer simulations. If laboratory-based there could be interaction with equipment. If the problem is seen to be authentic in the context of the course and engineering there can then be the development of an identity as students fell they are working as an engineer. The solving of a technical problem, resolving an issue or developing a design can all help develop the identity and belief they are undertaking relevant engineering-type activities. The setting of the challenge will need to be level appropriate. If the challenge is too much, or needs too much time, there can the anxiety related to the work, as Hendry et al have noted [122]. This could lead to a lack of engagement with the exercise and correspondingly a lack of learning, challenging the idea that the student can become an able engineer. This can be damaging to the creation of the idea that they are developing as an engineering student and feed self-doubt. Conversely, appropriate guidance and success can lead to the building of confidence and development of the identity as a future engineer. Perhaps one final point to consider with identity is that a medical case study is easily adaptable to a PBL problem and a solving such a case study is arguably like the day-to-day work of a physician. The translation of this type of encapsulated problem solving may not be as easily done in engineering, where design, construction and improvement may take a longer time. So PBL problems may be briefer in scope than in typical professional engineering work [208].

Another theoretical aspect to consider with PBL is the Kolb learning cycle. The problem and its solution can become the experience and the process of solving the problem can lead the student around the cycle, involving reflective observation on problem and its solution; abstract conceptualization with the discovered material and the problem solution; and active experimentation with regards to how to go about solving future problems. If there are a number of problems in a particular PBL course then the related moving through this cycle per problem can help develop the problem solving and learning skills. Encouragement of reflection by the tutor, as Hendry et al mentions [122], can be important for the students. Increasing the student's regularity of reflection on the problem-solving process could aid a more metacognitive development of the understanding of the approach to problem solving. Students will also reflect when they see in the later discussion stage of PBL how other students in the group approached their individual collection of information and problem solving.

5.3 CDIO

Another system approach to engineering education that has been developing since the 1990s is the CDIO initiative. CDIO is an acronym for Conceive, Design, Implement and Operate and represents the functional areas of contemporary engineering work [55]. The originating institution was MIT and unlike PBL this approach to education came out from the engineering school. The motivating force was to provide a new approach to educating engineering students that met the needs of industry and prepared students for the current range of engineering work. Joining MIT in this initiative initially were Chalmer's University, KTH Royal Institute of Technology and Linköping University [83]. The aim of the work was to develop an approach to educating engineering students such that:

> "Graduating engineers should be able to conceive-design-implement-operate complex value-added engineering systems in a modern team-based environment." [55]

Arising out from this challenge was a CDIO Syllabus [55]. This report [55] outlines the development of the syllabus, its creation of a set of goals for undergraduate engineering education, including showing how it meets the expectations of a number of requirements, such as the ABET 2000 accreditation requirements and Boeing's desired attributes of an engineer. This latter criteria set from Boeing shows the commitment of CDIO to create a framework for engineering education that produces undergraduates that are ready to meet current industry needs. The CDIO syllabus is split into sections and tiered into four levels, with the lowest level at an implementation level. At the top layer are [55].

1. Technical Knowledge and Reasoning.
2. Personal and Professional Skills and Attributes.
3. Interpersonal Skills: Teamwork and Communications.
4. Conceiving, Designing, Implementing and Operating Systems in the Enterprise of Societal Context.

A level down in Section 2 are the sub-sections:

2.1 Engineering reasoning and problem solving.

2.2 Experimentation and knowledge discovery.

2.3 System thinking.

2.4 Personal skills and attributes.

2.5 Professional skills and attitudes.

The detail gets finer, a level down for each of these lists a further set of sub-sections are listed. For example, for 2.2 experiment there are four listed:

2.2.1 Hypothesis formulation.

2.2.2 Survey of print and electronic literature.

2.2.3 Experimental inquiry.

2.2.4 Hypothesis test and defense.

Ultimately, the appendix of the report suggests implementation approaches; for example 2.2.3 Experimental inquiry includes: *The precautions when humans are used in experiments*, and *Experiment construction* amongst others [55]. As can be seen there is considerable detail and guidance in the syllabus structure, which has been used by over 100 programs around the world [56]. As of the time of writing the syllabus is at version 2.0 [56].

Following from the syllabus came a standard in 2004. This arose as a response as to how CDIO programs could be recognized CDIO Council, 2020. There are 12 standards that incorporate CDIO, which are shown in Table 5.3

Each of these standards has a description, a rationale and a rubric for self assessment. The assessment part allows the institution to evaluate where they currently reside with the standard, which is set on a six point scale (from 0 to 5). As can be seen in Table 5.3 there are a number of standards that directly connect to experiential learning, 5: Design-Implementation experiences, 7: Integrated Learning Experiences and 8: Active Learning. As well, 6: Engineering learning workspaces, which can facilitate experiential learning. Project-based learning is also encouraged by CDIO and on the CDIO website there are provided project learning modules. These currently include: a flight

TABLE 5.3
CDIO standards [67].

Standard number	Standard topic
1	The Context
2	Learning Outcomes
3	Integrated Curriculum
4	Introduction to Engineering
5	Design-Implement Experiences
6	Engineering Learning Workspaces
7	Integrated Learning Experiences
8	Active Learning
9	Enhancement of Faculty Competence
10	Enhancement of Faculty Teaching Competence
11	Learning Assessment
12	Program Evaluation

vehicle engineering project; a NASA rocket project and a skyscraper project [58].

It can be seen that CDIO it is different to the previously discussed approach of PBL as it is much more of a holistic approach to the syllabus, however it does not exclude the use of PBL within its structure, or other forms of experiential learning. A comparison of PBL and CDIO has been done by Edstöm and Kolmos [83] which looks at the histories, communities, definitions, curriculum design, relation to disciplines, engineering project and change strategies. They note that both approaches can be seen "as compatible and mutually reinforcing" [83].

There are annual conferences on CDIO and there are reports of the use of experiential learning and CDIO. One report illustrates how the CDIO helped to guide the development of a laboratory space with full size structural systems that could be used to help instruct and educate students in structural aspects of civil engineering [197]. This was an attempt to move away from the use of traditional lectures and scale models of systems for experiments. The result of a 'conceive-design-build' process produced a learning space that had structural elements of the actual space uncovered so students could see the different parts being used to make the structure. Further structural measurement equipment was installed that could log data and allow students to examine and study the data from a space and structure that they could see and use. This became a space for project work and interactive lectures where demonstrations and experiments could be made [197].

A paper that shows the value of the CDIO Syllabus is reported by Bertoni and Bertoni [28], who used the syllabus to categorize the key learning outcomes of students following a team project. They used the second and third levels of the CDIO Syllabus 2.0 to categorize the responses from students who had undertaken a team project and who had been asked to reflect on three key 'lessons learned' from the project. The question posed to the students was

> "Can you list three key lessons learned during the project work that you would share with future students in the course?" [28]

The student responses were mapped to the Syllabus 2.0 outcomes down to its third level and up to three syllabus outcomes were used per lesson learned. In this particular study the responses were from masters level students but the process is applicable to undergraduates. The course, on 'Value Innovation', from which the student's opinions were collected from, was a course that was focused on the development of innovative products of value and used design thinking. Students were from Mechanical Engineering, Industrial Engineering and Management, and Sustainable Product-Service Systems Innovation masters programs. Cross-disciplinary teams of 4 to 6 students were given a design brief and work was undertake at partner company locations. So the learning had a 'real world' aspect to it. The results showed lessons learned could be found in all four sections of the CDIO Syllabus (listed earlier). The distribution at the second level could then be collected and it was found that 80.1% had a lesson learned in the 2.4 *Attitude, Thought and Learning* sub-section. The next three highest was 3.2 *Communications* (78.7%), *Designing* (63.8%) and *Teamwork* (62.4%). Following analysis at level 2, the next level can then be examined. With the top three 4.3.1 *Understanding needs and setting goals* (56%), 3.1.2 *Team operation* (50.4%) and *Time and resource management* (43.3%) [28]. The investigators also looked at the differences in the lessons learned between the different programs. This study shows how a learning experience can be carefully examined by the well designed CDIO Syllabus. This syllabus provides a way to consider the lessons learned by students using a thorough and standardized set of categories.

5.4 Project-Based Learning

Project-based learning is often connected to problem-based learning. There can even be confusion as the acronym is the same. (For clarity in this chapter we will use PBL for problem-based learning and not use any acronym for project-based learning). Sometimes to acknowledge the link as well as the differences the term 'problem- and project-based learning' or PPBL [47] is used, sometimes the term project-orientated learning (POL) is adopted. Regardless of the name, at the heart of project-based learning is usually a team-based task to produce something, and the learning is associated with the experience of the creation. Having already covered problem-based learning earlier in this chapter we will outline in this section project-based learning and compare it to problem-based learning. In engineering there has been a long tradition in using projects in the curriculum, especially at the capstone stage, so the next chapter is focused on looking at the use of projects in learning. Therefore, details and example of projects will be kept for a later chapter.

A project is usually an assigned task with a specific deliverable. That deliverable could be a report (or similar), the construction of an item, or both. Because of the deliverable, the focus of the project is something specific and the learning related to the project is the application and development of prior knowledge to produce the required output. This is often why projects become a capstone activity. Projects can be done individually or in teams. The advantage of a team is that the task can be scaled (there are more individuals working on it) and there is also a need to learn and apply team-based working and communication skills. For engineers this allows the practice of skills that will be needed and used as a professional. Projects could become more of a design focus and sometimes this may be known as design-based learning [209]. If we compare the defining of the work involved in a project to that of problem-based learning we can see that the project will be usually better defined at the outset. Problems in PBL are intentionally more ill-defined to encourage the students to investigate and define the problem. Where there is ambiguity in a project is often found to be in the approach to the solution. This is where alternate approaches to the solution can be explored and design decisions need to be made.

It can take a significant amount of time for students to work on a project, so often a single project will last one or perhaps two semesters. This is unlike problem-based learning where a problem can last typically a week and a number of projects can be included in a term. There is a further difference in that a project can have certain stages or milestones, which can guide the project over the longer time period. Whereas the problem-based work is so short and focused on knowledge and information gathering that there will be little need to have milestones, though as noted in Section 5.2 and shown in Table 5.1 there is can be seven distinct stages in PBL.

The role of the students in a team, as well as the instructor, can also be different between the two approaches. In a project, students often take on different roles and undertake different aspects of the design. Whereas with problem-based learning each individual will be undertaking their own research and knowledge investigation into the required topics the group has identified. Teamwork may then become a little different, with projects often having individuals having their part of the project connected and even dependent on another's work. With a project the instructor may be an advisor and a guide, whereas in a problem-based learning situation they may be more of a facilitator and observer, ensuring the right learning objectives are being met. Table 5.4 shows a collection of some of the key elements and aspects of the two different types of learning. This is a coarse comparison and there may be differences with specific implementations of the types of learning, but this does illustrate some key similarities and differences.

TABLE 5.4
Comparison of different aspect of project-based learning with PBL.

Aspect	Project based	Problem based
Duration	Often single project in a course	Multiple problems in a course
Grouping	Often team based, but could be solo	Team based
Individual roles	Individuals have different roles	Individuals research same problem/topic
Knowledge	Used to apply and develop prior knowledge	Construct new knowledge
Challenge specificity	Project can be specific (approaches may be vague)	Problem could be ill-defined
Output	A solution, product and report	Knowledge through seeking a problem solution
Instructor role	Advise and provide guidance	Facilitator. Ensures learning objectives met
Objectives	Design, problem solve, teamwork and communication	Develop knowledge, problem solve, communication and teamwork
Experience	Design, development & construction	Investigation, research and problem solving

Kolmos provided an early examination of the difference between PBL and project-based learning [151]. Donnelly and Fitzmaurice provide another comparison discussion and also focused on differences in the roles of the students and the tutor between the two approaches [77]. In this latter work the authors do illustrate that not having the tutor as an 'expert' but instead as an 'adviser' the students gain a level of independence. This independence can allow the self-directed learning to grow. Another point the authors make is the need for an induction process to whichever approach (either project or problem-based learning) is adopted. This can involve explaining the process, having past graduates talk about the advantages, use of guest speakers such as employers of past students who used PBL, and the approach to assessment, amongst others [77]. As the learning process can be a departure from the perhaps the traditional lecture-based course, there needs to be some form of familiarization and reassurance of the approach to be used for the students.

Considering learning frameworks for a moment, if the learning is to be viewed through a situated learning framework then there are differences between project-based learning and problem-based learning due to the differences in the role of the instructor, the roles within the team, the activity (project or problem) and the time dedicated to the activity. The location of the work can also be a factor with the possibility that project work may be undertaken in a laboratory setting and with different equipment. As mentioned in the earlier section on PBL, it is more likely that a constructivist approach may be considered for the case of problem-based learning.

5.5 Challenge-Based Learning

Having already mentioned problem- and project-based learning, it is worth noting there is another approach to learning called challenge based learning (CBL). Here a group of students tackle a chosen real-world problem and apply their disciplinary knowledge and approaches to try help solve or resolve the particular challenge. Apple developed the idea of challenge-based learning and proposed it in 2008 as part of its Apple Classrooms of Tomorrow – Today ($ACOT^2$) [54].

There have been earlier uses of the term 'challenge based' with activities before 2008 for example by Piironen et al [212], who created a product development challenge project in embedded systems for second year IT students. This started in 2004 and followed a project-based learning approach. The Apple model of challenge-based learning was developed with a focus for high schools and intends for students to work collaboratively to help towards solving a challenge and to share their experience. There are recommended attributes to this type of learning including working on a universal problem with 'multiple points of entry', finding 'connection with multiple disciplines', requiring students to do rather than just learn and the use of Internet-based tools to collaborate and organize [54]. The Apple framework has recommended stages which are:

> "Setting up a collaborative environment
> Introduction
> Team formation
> Assessment
> Guiding questions
> Guiding activities and guiding resources
> Prototype/testing
> Implement
> Assess
> Reflection/Documentation
> Publish
> Ongoing informative assessment" [54]

An aim of the process is to help students develop 21st century skills. There is a guiding organization with associated information on the Internet at challengebasedlearning.org [60]. They provide a three phase framework; Engage, Investigate and Act, with an overarching stage of; Reflect, Document and Share. This initiative is targeted for high school students, which is exciting to see that students entering engineering programs at universities may have encountered an approach like challenge-based learning.

At the university level there has been a major focus on challenge-based learning at Tecnológico de Monterrey, including in their engineering programs. CBL has been included as part of a number of changes to the curriculum approach as part of their Tec21 Education Model [183]. Besides CBL there are three other pillars: flexibility and customization, inspiring professors and memorable experiences [188]. This new model has been operating at Monterrey since 2013 and aims to help students tackle future global challenges, such those related to climate change and a need for sustainability [183]. The aim of these changes is to help develop discipline competences as well as process or 'soft skills'. The use of CBL is to have the students engage with a real problem to solve and unlike PBL the aim is not to necessarily solve the problem, but instead to learn from the process of trying to solve the problem/challenge. In moving to this type of model of learning there was a competency focus in the curriculum rather than a subject, with the challenges being the core for the design and a move to modularization rather than separate courses [183]. Faculty were trained by Monterrey's Center for Educational Development and Innovation to help facilitate this new model and there was an adoption of Internet-based tools, such as a learning management system, video conferencing and communication tools, and a program/curriculum progression tool. The faculty role changed in the CBL model from the lecturer and expert who transfers knowledge to the students, to being an advisor and co-learner in the challenge. To facilitate the flexibility pillar the students could join designated flexible, interactive and technology (FIT) courses from anyplace using the online video conferencing tools. There was also the involvement with external 'training partners', who would be interested in the challenge and provide further expertise and knowledge. Evaluation routes within the CBL program could vary and were reported to include the following instruments; surveys, preparation of draft resolution, student interviews, observation scales, checklists, rubrics, prototypes, tests, statements of the proposed solution, videos and debates [183].

Membrillo-Hernández and colleagues at Tecnológico de Monterrey examined the Mechanical and Mechatronics Engineering program, as well as the Bioengineering program [183]. Their study looked at the views of the students and faculty, as well as the performance of the students in the CBL courses. They found that in the Mechanical and Mechatronics course the CBL students had an average learning higher than students undertaking a traditional course. As well they reported that those undertaking the online FIT courses had on average a better solution to the challenges to those who undertook the CBL face-to-face. Regarding the evaluation students were in favour (78%) that the clearest evaluation was through the rubrics and checklists [183]. Perhaps not surprising, as these can clearly lay out what is required to be achieved and at what level. The faculty were also surveyed after training with the educational development centre for satisfaction and feedback. The more popular feedback highlighted concerns on: the changing of established teaching habits, the proposed evaluation schemes needing practice, and that other faculty member's expertise may be needed with the challenge, as aspects of it may go beyond their own expertise. There were also concerns about students not acquiring the breadth of knowledge of the subject, as well as not being prepared for that type of learning. Despite these faculty concerns it was reported that students gained improved learning through the CBL approach, compared to traditional face-to-face methods.

The success of CBL at Tecnológico de Monterrey and the interest from high schools indicates that this form of learning may grow in the future.

5.6 Chapter Summary

We can see in this chapter there are some well-developed different approaches to learning that are being championed by a number of universities and institutions around the world. Sometimes these approaches are linked and even blurred, such a problem-based learning and CDIO could have overlap. But they can be quite distinct in execution. For example, problem-based learning is focused on research and knowledge acquisition to solve a problem, whereas challenge-based inquiry is more about learning through the process of trying to solve a real and challenging problem (which may not even be solved). All move the learning process heavily onto the students through the process of interaction within the approach, whereas a lecture-based approach has the students as passive receivers. Each student will have their own experience in the learning process, rather than experience the same lecture.

Each approach has key features and characteristics. CDIO provides the most detailed and structured approach, that can extend throughout the curriculum. For any faculty member, department or school wanting to implement these approaches there is supporting literature. This can show that they can be successful in helping students learning over the traditional lecture approach.

A question can be raised about the effectiveness of these methods. For PBL, which is perhaps the oldest of these approaches, it is has been an area of discussion in medical education which chronicled in Norman and Schmidt's 2016 paper [202] when they revisit a discussion that occurred in 2000 [201] and [66]. They do acknowledge there is likely the PBL offers "minimal difference" to the other established teaching methods [201]. But they do highlight

> "... PBL does provide a more challenging, motivating and enjoyable approach to education. That may be a sufficient raison d'être, providing the cost of implementation is not too great." [201]

The authors had advocated more theory-based research into PBL. They do note in their 2016 paper "Problem-based learning is not the panacea promised by its advocates" [202] and still supportive of their earlier paper's work. So despite the lack of evidence of significant difference in student knowledge, the methods can provide motivation, enjoyment and the ability to develop team and communication skills. A study on PBL effectiveness in engineering education by Galand et al indicates there can be improvement through PBL and project-based learning compared to a traditional curriculum [104]. They conclude the results

> "... suggest that students following this PBL curriculum, whatever their previous level of achievement, developed new skills compared with students following the previous traditional curriculum." [104]

A study of the effectiveness of PBL with library instruction for first year engineering students have also been done and found to be effective and more interesting than lecture-based learning, along with developing group work, communication and discovery skills [130].

Here we have focused on the efficacy of PBL as an example approach to illustrate that new approaches may show some improvement over lecture-based learning. With care in implementation they can be motivating and as effective or better than traditional courses. They come with the advantage of developing associated skills of teamwork and communications. Students will be learning from others and may be motivated to solve the problems or participate in projects.

A follow-up question can be raised; are there challenges to the use of these techniques? One issue can be the time it takes for faculty to develop, prepare and implement the problems, or projects for these techniques. This could be coupled with an apprehension to move away from a lecture based technique which may be familiar and yields results. There is also the time of students to account for. If considerable learning is needed outside the scheduled class time that can have an impact on student who may be balancing a number of courses. In the Monterrey CBL study the authors report that 71% of students noted frustration at not solving the challenge problem due to lack of time [183]. It is clear that to adopt these approaches there has to be a decision to spend the time and maybe to supply support, to ensure that faculty can design the approaches to be taken and the assessment. This invested time does seem to offer reward and the increased focus on these different approaches and the global adoption shows a commitment to how these approaches can benefit an engineering student's learning. There seems also strong commitment to form communities of instructors and to share knowledge through conferences and journal publications.

Having mentioned project-based learning, in the this chapter, we will now, in the next chapter, look in further detail on the project in engineering courses. This has been an established and long-term alternative approach to lectures, to help students learn.

6

Projects

6.1 Introduction

Perhaps one of the most significant forms of experiential learning that almost all undergraduate engineering students undertake is a project. This is usually encountered towards the end of the program and is often called a final year project, capstone project or it may be a project within a capstone design course. Often a multi-term project this may be an individual research project, but more commonly it is a group project that involves a design and build. Either way, it can draw on any the information and knowledge covered over the previous years of study, as well as the skill of learning new material and fitting it with existing knowledge. However, project courses can be found at other stages in degree programs. They are included in first year and upwards, to enhance learning through a different way to the more conventional lecture- and laboratory-based courses.

Compared to classroom-based courses or laboratories, projects can be a significantly different learning experience. In a classroom topics of a course can change weekly or each session. Similarly, laboratories are practical experiences that are completed in one or two schedule laboratory exercise. For a project, the challenge is larger in scale and can take a number of weeks, a term, or two or three terms as is usually the case for capstone projects. Correspondingly the task undertaken by the student is larger and involves a more significant examination of a technical problem. More often than not the there is a significant amount of design involved and it is done within a group. Students have ownership of their project and are required to plan and execute their work. Instructors usually guide and do not directly lead the students. The experience the students then have becomes more like a piece of professional work and assessments often mimic the type of deliverables an engineer would have to provide, such as proposals, design reports and presentations. There is none of the regular written examinations, encountered in many other course. Instead there is the need to draw on the technical knowledge from other courses, as well as the learning from the communication courses to help with their writing and presentation skills. The duration and depth of work can

DOI: 10.1201/9781003007159-6

make the project a significant and memorable learning experience. No doubt most of you reading this will have memories of your undergraduate capstone project, but perhaps less recall on some of your other term long courses. The detailed work, the struggle with open-ended problems and having to guide oneself through the work can provide a deeper and more memorable learning experience that has impact. In the case of the capstone project it is the need to draw together different knowledge and skills, whether technical, organizational and communication based.

Projects can provide direct hands-on learning. This can require extra time, resources and planning. A fundamental question can be 'what are the advantages to this approach to learning?'. In this chapter we will look at the projects broadly. This will include looking at capstone projects, multidisciplinary approaches, assessments in project work and non-capstone projects. We will start with some theoretical considerations.

6.2 Theoretical Consideration of Projects

Before we enter details of approaches to project work let us consider some of the theoretical ways of considering learning through project work. There will be more of an emphasis on a situative framework, as used throughout this book.

If we examine the project from a theoretical perspective what can we understand about the project process? First, let us consider capstone or final stage projects. These are intended to be a major cumulative piece of work that in many ways mimics professional work, which a graduate may soon be entering. From a situative learning perspective the learning experience is drawing the student deeper into the community. There is an responsibility for individual work, under supervision, and often done within a team. This can lead to a development of the individual identity. The project activities, involving planning, design and construction also help to reinforce the identity, but of course the experience of the activities leads to learning. Within the design there is the use of tools and equipment, a development of expertise in the project area and a learning of related jargon. Even if the project involves a team, each student will have their own different path of learning from the start through to the finish of the project. Figure 6.1 shows a breakdown of the situative framework into some of the elements in learning through project work. The activities shows the key stages, from defining the problem to be solved, the acquiring of parts and equipment, the design and construction phases up to reporting on the completed work. The mediation elements can involve language including new terms, especially if in a multidisciplinary project, where the terms, acronyms and definitions need to be understood from other disciplines to facilitate effective communication. The specifications, equipment and analysis

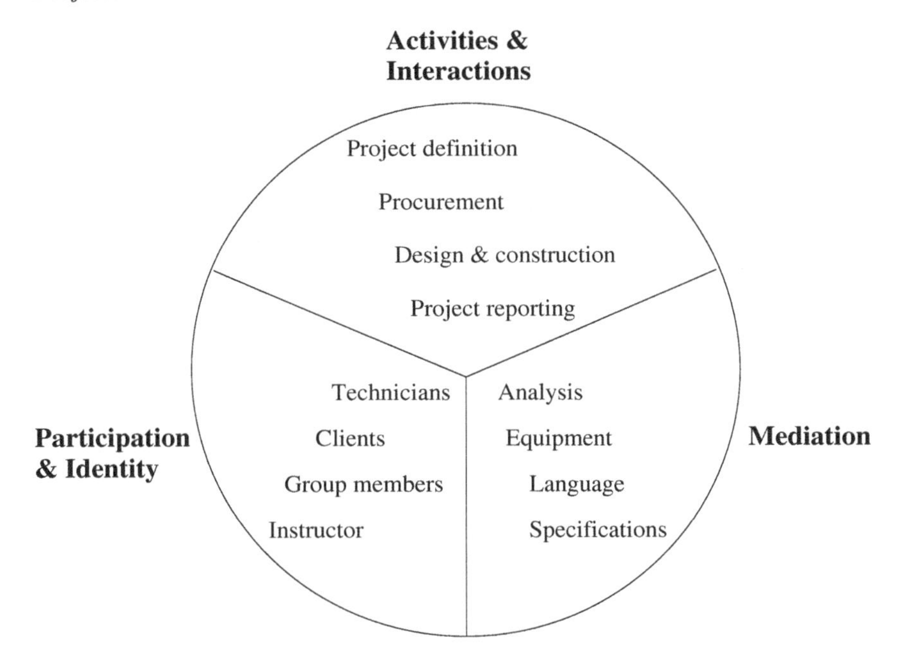

FIGURE 6.1
A situative model of learning through a project focused on three areas of interaction, mediation and identity.

techniques, including diagrams, are also mediating objects. In the 'participation and identity' section the various individuals whom a project student can interact with are detailed. Guidance is provided by the instructor who may take a manager or lead engineering-type role. Other group members can have different roles and responsibilities and that can define to some degree the identity of each individual in a team. The relative knowledge may also become a differentiator too and so student team members may help each other. There may be clients involved who have requirements and a need for support, plus assistance can be provided by technicians.

If we consider projects that are not at a final stage, but an earlier year in the degree program, including first year, we still can see that these can be important stages in a situated learning experience. Compared to a capstone project, there is likely more guidance from the instructor and the work can be on a smaller scale. The individual experience will still be important and the opportunity to have individual advice from an instructor or a TA will allow face-to-face instruction and learning to occur. Sometimes laboratory experiments are prescriptive of approaches or equipment to be used. In a project often the equipment and resources of one or more laboratories are available to a student. The process of learning how to select equipment, components and materials can be an important stage of development for the student. Such selection and choosing can be a regular part of a professional's work, but often something not done in a class environment on a typical course. Again the learning environment is encouraging a student to make decisions like a professional and that is developing the identity in a situated learning perspective. We have already mentioned the interaction of students with faculty, but the interaction with peers who have responsibility on other aspects of a group project can become important. Now, we have peer-to-peer technical communication on decisions and approaches. The resolving of different perspectives on approaches and views leading to decisions, has to be done. In a student and instructor discussion the instructor's experience will have considerable authority. In a peer discussion there is stronger equity in experience and so the resolving of differences in opinion has to be dealt with in an effective and evaluated way. This too can draw students into valuable learning experiences that help them develop and progress in the situated learning community.

If we take a cognitive framework approach to projects we can see a capstone project is intended to draw from the cumulative experience of the way problems has been solved in their previous courses. Models of understanding of problems and problem solving can be applied in the project. The experience of working in a laboratory helps in understanding how project stages could be broken into laboratory-like exercises and measurements and tests can be done. Students will be motivated to undertake work they are interested in and feel that they understand.

Considering projects from the perspective of the Kolb learning cycle we can see that they can be a learning experience that allows moving around the cycle a number of times, because of the longevity of the course. As well it can cause some dwelling at the various stages around the cycle. If we consider capstone projects the initial stages may involve significant time at the reflective observation stage (after the initial brief and specification), as well as abstract conceptualization when relating potential work and the design with what has been learned. Experimentation and experience can follow as work gets underway. Projects at other stages of a program, for example first year, may start strongly with an experience, perhaps a guided construction. That experience then starts the cycle and encourages reflection.

The use of Kolb and a framework can provide a valuable way to consider the learning that is ongoing in a project. Within my own teaching in the COVID pandemic colleagues and myself had to adapt our project courses and some of us used a combination of situative learning framework and Kolb to look at how project teaching was different when students were working in their own residences and connecting through online tools [251]. This is detailed in 11.4, but an example will be given here. In a third-year project I facilitate for electrical engineering students an initial group activity is to assemble a microcontroller board from a supplied circuit board and the components. This is a good way to get the group working closely together to solder the board, program the microcontroller and then test. Problems encountered could be hardware or software related and so it requires the experience in assembling and testing the board. Observing and reflecting on the performance and if there are problems considering where the problem may be. In the pandemic situation with students away from the lab, the initial exercise changed to the evaluation and selection of an already constructed microcontroller boards. Now the exercise required the students to consider their application, the specifications and evaluate microcontrollers for that use. So, rather than the experience of building and debugging a board, the reflection and evaluation of an available board rooted the initial exercise in the reflective observation and abstract conceptualization stages.

One additional point to make when considering the Kolb learning cycle with project work comes with the related learning styles from Kolb and the way individuals in a group work together. As mentioned in 2.6.1 many engineering students have learning styles that connect to active experimentation. Either leaning towards more theoretical consideration, involving more abstract concptualization, so in the *converger* quadrant. Or, alternatively more towards concrete experience, so in the *accommodator* quadrant. However, if a student has a learning style that favours the other quadrants, or a different quadrant to the majority of the group, this could lead to a significantly different style of working and potentially leading to conflicts. Diversity of ideas and approaches are valuable, but a project supervisor may want to be observant when different learning style are leading to frustration and potential conflict between inexperienced students. Here is an opportunity to learn about professionalism and respect for others.

Having considered some theoretical aspects, let us look at some types and aspects of projects.

6.3 Capstone Projects

The project, sometimes called a design course, in the final year of study is a key piece of experiential learning for engineering students. Encouraged through accreditation organizations and the move in recent decades to increase the amount of design work a student experiences. A survey of capstone design courses in the US has been repeated three times at approximately 10-year intervals [128, 129, 261]. The 2015 survey [129] reports data on how long capstone design courses have been in existence, see Figure 6.2. Although 6% less than five years old, while 7% are 50 years or older, the data shows that 71% of the respondents are between 10 and 39 years old. Which covers the period of 1976 to 2005. This rise of the capstone design course in the US in that period is likely due to the ABET accreditation requirement change for a design course to be included in programs. This change is noted in Floyd et al's section on the "Third Major Shift: Renewed Emphasis on Design" in their paper "Five major shifts in 100 years of engineering education" [101].

They also note that these courses were developed to address concerns that students were not as prepared as they could be for work in industry. Outside of the US were similar moves towards including more design. In 1993 the eighth recommendation from the Canadian Academy of Engineering's report on Engineering Education in Canadian Universities was:

> "The curriculum should include at least one opportunity to undertake a major design task. The selection of this major design should be such as to emphasize a holistic approach. " [84]

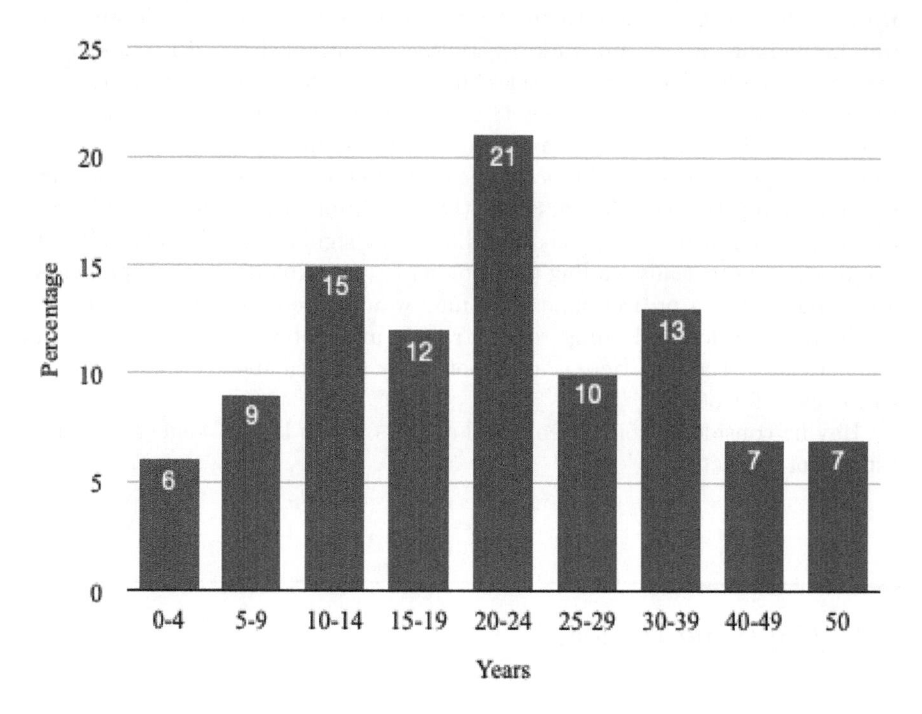

FIGURE 6.2
Number of years capstone design courses have been run in US institutions, using data from the 2015 decennial survey of capstone design courses [129].

Moving into the 21st century the Royal Academy of Engineering in the UK published a report in 2006 looking at the education of engineers in the 21st century by talking with industry. From the findings on what was a desired by industry in an engineering graduate at the top was practical application followed by theoretical understanding [248], p. 54. It was reported that industry desired from graduates more realistic 'real world' perspectives, and capabilities with practical aspects as well as theoretical aspects. On how to deliver this the report provides some responses including:

> " 'Practical project work, preferably with an industrial partner is necessary' argued one respondent. 'More creative, practical and market led project work,' suggested another." [248]

Clearly there was a continuous belief over the last three or four decades in having courses that involved students in gaining design knowledge and experience. Motivation for this can be viewed as what industry requires as over that period there was significant technological change, including the growth of information technology and global manufacturing and trade.

Data from the three decennial US surveys on capstone design courses shows that there has been a move by engineering schools towards having a project with courses in parallel over the 30 thirty years (1994, 48%; 2005, 55% and 2015, 68%), whereas projects only were decreasing (1994, 26%; 2005, 21% and 2015, 13%) [129]. The other types of course offerings were classes followed by a project, classes only and other forms to including design. The classes only option was low and decreased to zero in 2015 (1994, 5%; 2005, 2% and 2015, 0%) [129]. Clearly the capstone project had become a key way of including design and open-ended problem solving, that would be needed by graduates in industry.

The project has evolved in many institutions over the years from being an individual student project to one where a group of students work together. In 1995 the first decennial capstone design course survey [261] found that 83% of engineering departments showed the course could be undertaken in departmental teams, whereas 32% could allow individual projects to be done individually (presumably some institutions allowed both). It is interesting to note that in this 1995 survey 21% of respondents allowed inter-departmental teams. We will discuss multidisciplinary projects in a later section.

The sources of the projects were examined in the survey and in 1994 the results were 59% *industry*, 58% *internally* and 15% *other*. In the subsequent two surveys there were more response options and the top three for both surveys were *industry/government* (2005, 71% and 2015, 80%), *faculty research* (2005, 46% and 2015, 53%) and *external competitions* (2005, 24% and 2015, 35%). We will explore engineering student competitions in Chapter 8. The project can provide a taste of what it is like to work on a project in industry or in a research group.

The open-ended nature of the project provides the student with a challenge. They will have to better define any vague aspects of the problem or the task required of them. Unlike in other courses there will a focus on self-direction and organization. That organization in team projects will require role definitions and work allocation, planning and scheduling. The focus of the topic maybe unfamiliar to the student(s) and this will require direct application of prior knowledge and learning new details. Team members will need to share knowledge and findings. Solution choice through the design will challenge and engage students to make key decisions. These elements provide opportunities for the development of life-long learning skills.

The cycle of a typical project starts with defining the problem and moves through research, development and testing, towards completing the project and final reporting of what has been achieved. Table 6.1 shows a typical breakdown of the possible stages of a team-based capstone project stages.

TABLE 6.1
List of potential activities in a capstone project at different stages of time.

Stage	Technical activities	Organizational activities
Early	Defining the problem	Getting to know team and organization
	Assessing scope of work	Planning and scheduling
	Undertaking research	Regular meetings
	Solution ideas, feasibility and selection	Decision-making
	Planning design	Parts procurement
Main	Solution development and construction	Progress reporting
	Design review	Review presentation and receiving feedback
	Solution redesign	Re-evaluating the schedule
	Integration of elements students have worked on	Planning for completion
Final	Final Testing	Drafting final report
	Demonstration and design evaluation	Finishing to schedule
	Documenting and reporting	Completing final report
	Final product handover	Final presentation

The technical and organizational activities may vary between projects and not all of the stages may be involved, but the basic processes for most projects are outlined. A resource to help guide students through the design process is a design workbook that offers note on the design process and templates for students to use [210].

The process outlined in Table 6.1 shows the overall progression followed. Atman et al has studied the design process cycle for students at different stages in their programs, as well as engineering practitioners [18]. This examination of the process was done using a verbal protocol analysis [17] of the design process taken by students and engineering practitioners when asked to design a children's play structure, in a three hour 'pencil and paper' design. By having the designers saying aloud what they were currently doing the verbal protocol approach could be applied and the time spent on various stages of the design process could be monitored. The stages of the design process were broken down into eight activities: *problem definition, gather information, generate ideas, modelling, feasibility analysis, evaluation and decision, communication.* The designers were from three different groups, freshmen engineering students (n = 26), senior engineering students (n = 24) and engineering practitioners (n = 19) where were from different disciplines. The quality of the solutions were also assessed which allowed the analysis of the design process taken for each group to be considered for three different levels of quality. A timeline of what design activity the different groups were at could be created for low, median and high-quality solutions. It is worth consulting the paper to look at the detailed graphs of the timelines for the different designer groups and quality levels. Some of the summary findings show that experts spend more time on problem scoping activities, which are the *problem definition* and *gather information* stages. Accordingly, experts gathered and requested more information than the students, including more attention being paid to accessibility, safety, budget and information on the location. As well, more time was spent by experts on project realization, that is the *decision* and *communication* stages. Quality of the final assessment of the design was not too different between the senior students and the experts. Though unexpectedly both were of higher quality than freshmen designs. The overall project quality was self-assessed by the designer and interestingly the experts rated their designs lower compared to the students. The authors of the study explain this could be due to experts "know they were working outside their domain of expertise in an artificial setting" [18]. One particular concluding point by the authors is that students would benefit if they know about the stage structure and the need to keep iterating through stages to help feed information into the design and project realization along the project timeline.

Professional skills are developed within these capstone projects. The previously mentioned capstone design course surveys provide insight into the top topics covered in these courses. In the 2015 survey the top five topics covered overall from respondent's projects were: *written communications*, *project planning and scheduling*, *oral communication*, *concept generation/selection* and *team building/teamwork* [129].

If we consider the Kolb learning cycle the project stages can ensure that there will be multiple passes around the learning cycle. As students experience different activities they will progress around the cycles a number of times. Some cycles will be small and at an individual level, such as getting some subsystem built and working. Others will be a large-scale activity that involves multiple team members, such as the integration and final testing activities. Figure 6.3 illustrates a typical example of a student embarking on their initial work in a project. The cycle involves what the task involves and what is needed (reflective observation), relating the work to previous knowledge from courses or the literature (abstract conceptualization) and then deciding on the steps to be taken and embarking on the first steps (active experimentation). The advantage of the Kolb cycle is to capture the thought-processing stages, so a more detailed and holistic consideration of the student's learning can be made, rather than looking at just the actions or outcomes at the end of a cycle.

If this is the first time a student has encountered a project then the learning associated with the scale of the work will be significant. Scheduling and planning become key element as progress through the stages can sometimes

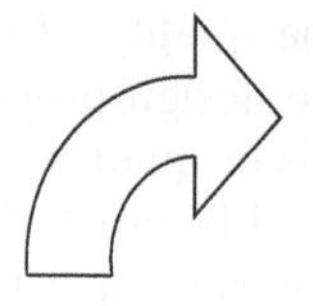

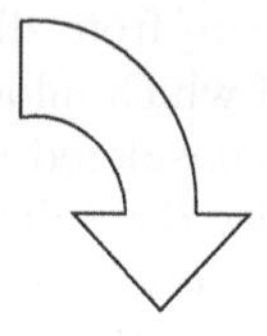

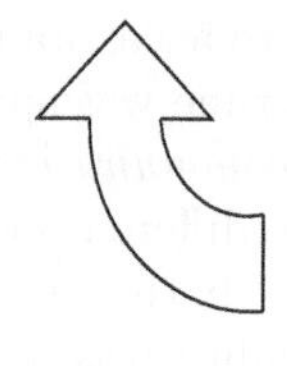
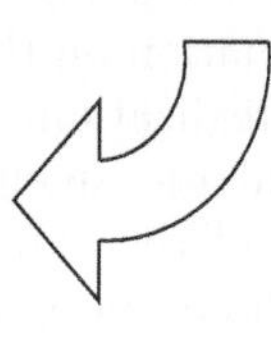

FIGURE 6.3
A Kolb learning cycle with an allocated project task.

move slower than many inexperienced students can appreciate. Likely a key reason for the inclusion of that as a topic in design courses (see the earlier mentioned top 5 topics). Many project supervisors will be familiar with the accelerated pace and work catch-up as the second or final term is entered. The student's learning experience of self and team organization and scheduling become important. This is partly the reason why there is an increase in including project work in programs *before* the final year. More about projects in the earlier years of engineering programs later.

Issues and problems encountered in the technical work can have impact on the project progress and consequently its schedule. Here is where previous experience in the laboratories will help. However, there may still need to be learning about problem solving, as inevitably, problems occur and solutions are sought. This is perhaps when reflection can be valuable.

We have already outlined project work from a situative learning framework in Section 6.2, but it is worth revisiting for capstone projects. The project usually has a community around it, which varies in size. We can consider the community members, activities interactions and mediating artefacts, as previously mentioned and detailed in Figure 6.1. The community comprises an number of different participants, from project supervisors, other students, technicians and possibly industry representatives. Guidance is provided by the instructor who takes a supervisory role. If the project is multidisciplinary (more about that in the next section), there may be supervisors who are from other departments and this could make the community quite large. This may be rare chance to collaborate with a supervisors and student from another discipline of engineering. Other student group members can have different assigned responsibilities and that can aid the defining the identity of individuals in a team.

The activities show some of the key stages, from defining the problem to be solved, the acquiring of parts and equipment, the design and construction phases up to reporting. The stages a project goes through has already been mentioned, see Table 6.1, and involve activities and the mediating objects. For example at the start there will be some form of requirement or specifications to a design. Proposals are then drawn up and scheduling is tackled. The mediation elements can involve language including new terms, especially if in a multidisciplinary project, where the terms, acronyms and definitions from each discipline need to be understood to facilitate effective communication. The specifications, planning charts, equipment and analysis techniques, including diagrams, all become mediating objects. When using equipment, especially more specialized equipment, there may be more interactions with technicians. Similarly if something needs to be made, there may need to be drawings made to ensure a technician can successfully build what is needed. This may take a student beyond what has been normally encountered in a laboratory. Explaining what is required and sometimes clarifying can then become a new experience, drawing on technical and communication skills. CAD tools may need to be learned which can involve a significant amount of time to learn and use.

The project with its stages and cycle provides a rich experience. Using and extended knowledge from prior courses can challenge the student. Interacting with others, including fellow group members can exercise the student's communication and organizational skills. We will now consider multidisciplinary projects where the team members are from different engineering programs.

6.4 Multidisciplinary Projects

The 2005 survey of capstone design courses showed that 35% of those who responded (n = 444) had multidisciplinary courses, this was an increase over the 1994 survey which had 21% (n = 360) [128]. As only 2% of the 2005 responders indicated that they had 'class only' courses we can assume that the majority of the courses involved projects. In the 2015 survey report (n = 522) it was stated that "Just over half of 2015 survey respondents included faculty and/or students from at least two different disciplines in their capstone courses" [129]. From this series of three surveys done over the decades we can see that there has been a trend in capstone design projects to move towards having multidisciplinary capstone projects. What then is potentially causing this move towards multidisciplinary approaches to projects?

The advantages of multidisciplinary projects can be varied. The increase in the level of expertise from different departments can allow projects to grow beyond the confines of a discipline. For example, an autonomous airborne vehicle could have aerospace and mechanical engineering students working with electronics and computer engineering students. If either group had to work on their own there would be distinct shortcomings in knowledge and expertise, limiting the scope of any design. Together a whole vehicle can be designed.

The linkage of disciplines and the collaboration brings together another aspect of effective team working and communication across the disciplines. Talking peer to peer on technical aspects takes on a different aspect when the communication moves outside of a single department and is across disciplines. This is not to say the demarcations are precise. Sometimes students may have to deal with difference and other time with overlaps or different perspectives in knowledge, which is valuable communication experience. This collective of expertise working together on a joint project with responsibilities in designated areas, can more accurately reflect the type of experience that can be found when working in industry. A quantitative study by Hotaling *et al* examined the impact of a multidisciplinary design course as compared to a mono-disciplinary design course [127]. The multidisciplinary teams were composed of biomedical engineering students and mechanical engineering students at Georgia Institute of Technology. They found that the multidisciplinary team produced a better design solution, as measured by professionals from industry, compared to their counterparts from either of the individual departments. The average increase in the score of the multidisciplinary team's projects from the capstone design expo of the project was 10% to 15% beyond what the single discipline teams achieved. They also found that members from the multidisciplinary projects were more likely to find employment within seven months of graduation, compared to the biomedical engineering students that did a single discipline project.

What then are the elements in a multidisciplinary project that provides the added success and value? Conversely, what are the challenges and problems that come with such a project? Thigpen et al provided a model of a multidisciplinary capstone design project at the early part of the 21st century [258]. They reported the evolution of a capstone design project for mechanical engineering students to move from a single discipline project to one that drew in students from electrical engineering, as well as business and fine arts. The project, supported by a car manufacturer, then developed beyond just mechanical engineering to involve broader engineering design aspect as well as marketing. The objectives were to provide a course that "(1) provide students with insight into the industry work environment, (2) develop their professional and technical skills, (3) prepare students to work effectively in diverse multidisciplinary teams, and (4) to apply the curriculum to effective product design and manufacture" [258]. The industrial supporter, General Motors, could provide strong help with (1). It could be argued that (2) and (4) could be provided in a single discipline project, though the professional skills may be challenged more and the design may be broader in a multi-disciplinary project. Having students from four different units certainly would make objective (3) significant and a strong learning experience for a student. Considering the objectives as a whole we can see that a student having experienced a course would not only have a broader set of capstone experiences, but they maybe more attractive to industrial employers for being hired. Having been a manager in industry who hired new graduates I know the capstone project can be a focus of attention in a hiring interview. Some of the advantages mentioned include: exposure to practical aspects of engineering, working in a multidisciplinary teams, product design experience along with consideration of manufacturing and marketing, and overall a clear stepping stone from classroom work to the type of work in industry. Challenges reported were somewhat faculty focused, including the time and effort in coordination, ensuring the overall project and its elements were achievable and the faculty from the various departments were advocates for the student project [258]. My own experience with multidisciplinary projects has shown that these challenges do occur. The larger student teams, often encountered in multidisciplinary teams, who are in a range of programs can often have different timetable schedules and so initial stages of projects can be dealing with schedule organization for group meetings and work. Another challenge that can occur is some students may need to spend time filling in background knowledge, for example, electrical engineering students entering into an aerospace project may need to learn some principle of aircraft and a little about aerospace engineering. However, these challenges can become learning opportunities within the projects as organizational problems can be solved by students and peer learning can be utilized.

A study looking at the how peer evaluation differed between team projects in either a multidisciplinary project or a single discipline project was undertaken and reported in 2019 [215]. The study looked at four single disciplinary projects, as well as nine multidisciplinary projects at University of California Merced. Students were from the majors of bio-, environmental, materials and mechanical engineering. Some of the projects were proposed by industrial partners and students were supported by instructors and industry representatives. The structure of the course was centred around design-based learning. Across 13 teams 60 students were peer surveyed on their performance (contribution and skill) and team satisfaction. The results showed that in the multidisciplinary teams the students rated their peer's contributions and skills higher, compared to the single discipline project teams, especially in the areas of 'keeping the team on track' and 'expecting quality' [215]. When it came to the team satisfaction there was no significant difference between multi or single disciplinary teams. This study was viewed as a preliminary study and the differences and similarities between the two types of project team composition were to be investigated.

As a final point on advantages of the multidisciplinary projects is the interest in encouraging the development of the 'T-shaped' engineer [223, 264, 288]. The use of 'T' indicates the structure of the engineer's knowledge. The vertical stem representing depth of specific knowledge and the horizontal bar the breadth of knowledge. As opposed to an 'I' shaped engineer that is predominantly a specialist. Multidisciplinary projects can help in creating the 'T' knowledge range, including the skills needed for cross-discipline communication. This and the other benefits may be worth dealing with the reported disadvantages of organizational challenges and the need for enthusiastic faculty to support the projects.

6.5 Assessment of Project Work

Project work, like other courses has to be assessed. Because of the nature of the course does not lend itself well to written examinations or tests.

It is usual that any reports that were produced as part of the project become a key part of the assessment. The final report of the project does often become the main approach to the assessment. This report is the culminating description of the project investigation and the findings from the work. Here is where the core and most important work will be included. However, there are other reports along the way that can be considered too, such as initial proposals, individual design reports, interim reports, posters and similar. In team-based projects separating out the individual contributions, or integrating a cumulative group mark into the assessment of the individual has to be done. Sometimes the use of individual design reports can be used in differentiating individual performance from the group. These design reports could be a brief technical-based report focusing on a specific area that a student worked on. If the students are in a team then their specific contributions to the project can be separated and an design report of a specific part of their work can be detailed. There is benefit in these reports being written as the project progresses, rather than leaving to the end, as it allows consideration of work done and it allows experience of writing reports and to obtain feedback, before the final report has to be made.

Beyond the reports there is often the assessment of oral presentations. These may be done as the project is ongoing, for example in regular meetings or at specific stages, such as an end of first term update in a two-term project. Frequently oral presentations become part of the final presentation of the project, along with a report. In presentations there are sometimes opportunities for questions to be asked and the student to be examined. Of course, more formal oral examinations are possible, but due to the number of students in a project course the time needed for this to be done can be a logistical problem and even prohibitively long.

Another part of student assessment is the contribution to the project. Some students put in extra effort, and sometimes key effort to make a project successful. This is often recognized by something like a 'contribution' assessment. The challenge here for the assessor is knowing how much and the quality of the contribution. Project work is often away from the direct view of a supervising instructor. So the contribution can be based only on results or interactions in meetings with the student. Although there can be a reasonable measure of an individual's contribution to their tasks, it may not catch some aspects of leadership qualities and input to other team member's work. One way around this is to have peer assessment or some way of polling for the contribution of others. In a project course that I have coordinated, I have asked towards the end for all group member to privately rate the contribution of others in the team, as well as indicate who have provided leadership to the group (and they do not need to be a designated leader). The term 'leadership' is qualified to include those who have helped individuals with tasks and generally provided support when needed. In a group environment it is soon possible to identify key individuals in a project team, when all individual assessments from the group are collected. Another approach to looking at the contribution is to provide a total number of points and to ask the group as a whole to assess who contributed to the project and then to divide the allocated points in manner that reflects the effective contribution. Because of the open nature of this discussion and splitting, it may lead to disagreements but if accompanied with other private assessments there may be a clearer picture of who has contributed to the project. As well, the process may be a valuable reflection for individuals and the group as a whole.

Having mentioned disagreements within a group it is worth mentioning ways to preempt such issues, amongst other problems such as a lack of contribution at the outset of project work. To do this I have employed a group formation form that includes some guidance questions for the group to consider, see Table 6.2. Some relate to ways the group will operate and other form the start of a group discussion, so there is a basis for the way the members of the group will interact and work together. Other questions have come from experience of seeing how occasionally issues can arise within groups.

These questions provide a framework of cooperation and ensures some potential risks to good teamwork to be discussed at the time of the team

TABLE 6.2
Organizational questions for a group. Used to help the groups organization and reduce interpersonal issues.

1	How often the group will meet?
2	Is there a group leader?
3	How will all contribute to discussions?
4	How will decisions be recorded?
5	How will information be shared?
6	How will roles be respected?
7	How will the group handle poor time keeping (late arrival, early leaving, not turning up or not progressing work)?
9	What does the group consider to be characteristics of a good team member?

forming. Often with the challenge of the project ahead and the excitement of getting underway some of the ways of how the group will work can be overlooked. Many of the questions have come from seeing issues and wanting to help groups avoid them, as well as consider the issue. For example, the respecting of roles can be overlooked, or even the recording of decisions (often in group meeting minutes) had not been adequately done and there had then be misunderstandings. The question of time keeping is effective for those who allow others to move the project forward in major ways. From a situated learning perspective this is a member of the community who is not engaging and consequently not learning what is expected.

The above questions in Table 6.2 were created for a third year design project in an electrical engineering program. Laboratory time for the project was schedule and the project was over two terms. This project experience was then a precursor to the capstone project of the following year. Lessons learned in this project experience can then be carried over to the two-term capstone project.

Another way of assessing students towards the end of a project, beyond a final report, is to consider the completion and solution to the project's problem. This is an approach described by Laguette [156]. In a mechanical engineering capstone design project that spans three quarters (Fall, Winter and Spring) at UC Santa Barbara, Laguette reports that students teams were required to submit a preliminary Project Completion Requirement (PCR) at the end of the Fall quarter [156]. In this document the students document the specific outcomes from the project and the evidence that will be used to assess the completion. The final PCR is submitted one month before the end of the project. The instructor involved with the project must agree on the team's documented planned outcomes and specified evidence. The final project review must then address the stated outcomes. This approach is noteworthy as a final project report may implicitly include information on the project completion, but this process openly requires the students to consider what is to be done and then has to measure this from their own criteria. The paper [156] includes an appendix with a scoring rubric that breaks the marking into looking at the prototype completion, the testing done and the modelling and/or analysis. Included in the completion package are instruction and user manuals, diagrams, software or source code and information addressing safety around the project and related materials. This focus on what is entailed in the completion of the project, as specified by the students, provides a route to assessing the performance, but also provides a reinforcement of the planning the students must undertake. The included documentation requires the team to think in terms of a user and that ensures an alternative perspective from a designer and developer view.

In 2008 a US group of educators and researchers called the Transferable Integrated Design Engineering Education consortium (TIDEE) investigated assessment methods and instruments for assessing capstone design courses [107]. The resulting model for assessing the outcomes of such courses looks at

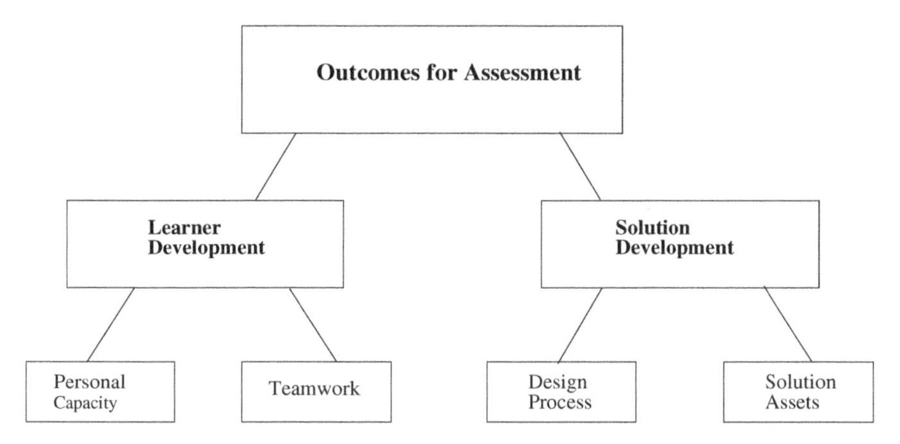

FIGURE 6.4
The Transferable Integrated Design Engineering Education consortium model of capstone design course assessment. This breaks down the elements of assessment areas. Based on the model in [107].

splitting the into two parts, *learner development* and *solution development*, Figure 6.4. For assessing *learner development*, individual performance and team working to be considered. Whilst the design processes employed and the produced results can comprise the *solution development*. Here the process of assessing the project work of a student can be split between the student's work in the project and what is achieved in the design. This splitting can help with assessment in a team project. When a group works on a project problem and the product becomes a combination of all the individual contributions there can be a challenge in how to assess individual contributions. If an assessment scheme is set up using both the learner development and solution development there are ways to extract individual work from group work. The assessor then needs to consider the relative weightings to the various assessment elements. Table 6.3 illustrates how various different assessment types that could be encountered in a project can connect to the two stages. In the table the two stages of learner development and solution development are further divided, according to the model in Figure 6.4. The demarcations between the two development areas are (given as columns in Table 6.3) are not always clearly defined and can depend on the requirements in the assessment description. So, sometimes the assessment can fall into both development areas. However, distinctions and categorization can be made. For example, the use of a reflection journal can be mostly focused on individual performance (though there maybe assessment of the individual's approach and contribution to a design), similarly an instruction or user manual is focused on the produced product and so solution development. An individual design report is based in both categories as the student is presenting an personal contribution, but the report will be detailing a design process and findings.

TABLE 6.3
Splitting of the assessments into the TIDEE model areas. Subdivision of 'Learner Development' and 'Solution Development' into the 'Team' or 'Personal' and 'Design' or 'Solution'.

Stage	Assessment	Learner Development	Solution Development
Early	Group organization	Team	
	Proposal/definition	Team	Solution
Main	Reflection journal	Personal	
	Individual design reports	Personal	Design
	Intermediate design review	Team	Design
	Progress reports	Personal	Solution
Final	Completing & demonstrating	Team	Design
	Final report		Solution
	Final design review	Team	Design
	Instruction/user manual		Solution

When coordinating a project or capstone design course the list in Table 6.3 can give a suggestion of a range that can be used. The assessments that fit within the project design course can be selected. Selection can draw from the two development categories, to create a balanced assessment across learner and solution development. For example, in a one term third year project course in electrical engineering that I have coordinated, the various assessments used are listed in Table 6.4. Besides the more group-based assessment, such as the project proposal, presentations and final design reports; individual work is required in the form of individual design reports and individual reflections. As well as these assessments the students are asked to discuss and rate the contributions of each other and present the percentage results (which should total 100%), in an appendix. These along with the instructor's observations can be used as a contribution element in the assessment.

The sequence of the assessments follows the order listed in Table 6.4. The groups have to decide on their own project, which they are going to design

TABLE 6.4
Assessment staging for a one-term project. Whether the assessment is to be completed as a group or an individual is indicated.

Stage	Assessment	Individual	Group
Early	Project proposal		X
Main	Individual reflections	X	
	Individual design report	X	
	Presentations		X
Final	Completing and demonstrating		X
	Final design report		X

and build, so the outline and planning are included in the proposal. At regular intervals along the term students add entries to their personal reflection journal. Towards the middle of the term there is an individual design report based on the work the student has been doing. At intervals there are presentations given so each group can see what the others are working on and their progress. One of the presentations is a reflection exercise for the group, on one important aspect learned by the group from the project. It is interesting to see how the groups can differ in what they consider is an important piece of learning. They can determine it to be a specific technical detail, through to aspects like teamwork. In the final laboratory session each team gives demonstrations of their project and the final report is due at term end.

The final point to note with project assessment is that a significant amount of assessment work can be done across the duration of the project and can be viewed as formative in nature. Each element of the assessments in Table 6.3 can help with the developing individual and team work. The assessments are authentic and help understand the design process.

6.6 Case Study

Activity: A project developing a cycle and walking path extension along the coastline of Auckland
University: University of Auckland, Auckland, New Zealand

Civil engineering students at the University of Auckland undertook a one term capstone project that looked at a current real-world engineering problem. In 2019, the second year the course had run, the project was to design a walking and cycling path from Orakei Basin to Tamaki in the city of Auckland. This was the final stage in a four part redevelopment of a pathway that links a suburban area to a harbour and provided some design challenges. It was 7 km long and ran along the coastline where there were boat sheds along the shoreline, so access for boats should not be impeded. The aim was to provide a path suitable for cyclists, so a minimum gradient and sweeping corners whenever possible, plus safe surfaces in wet weather. There were also geological conditions and minimizing of tree damage to consider too.

The project was a civil engineering problem that was underway at the time of the student project. Over 200 students were brought together into 27 teams of between 7 and 9 students. The groupings were done prior to the course starting by software and the faculty team, based upon submitted technical interests, experience and preferences by the students and the technical roles within the project. There were faculty advisors for:

- Construction management
- Environmental
- Geotechnical
- Hydrology
- Structural engineering
- Transport engineering

Information from the public domain on the project was collected and provided to the students, so they could have access to a wide range of information, including feedback from public consultation. The project was local so students were taken on a site visit and it was accessible if the students wanted to return to the area. Further information was provided by an engineer and the project manager from the company that had won the contract for the project. In a guest lecture both details of the project and the steps and stages of the project were discussed. The students were gaining deeper insight into a real project that they had already had detailed knowledge through their own design. The opportunity to compare their approach to that of the company provided an important learning opportunity. Should any other specific information be required then there was a stated process for requests for further information, which were directed to a faculty member.

The course length was one semester which consisted of twelve weeks with two weeks break in the middle. This course was done in addition to a required two semester research project.

Support from twelve faculty (two per specialist area) was provided with scheduled weekly meetings of groups. Faculty with different technical expertise would move between rooms where students were in their groups and opportunities for informal question and answer sessions as well as discussions were possible. Using these 'roving' specialists provided a degree of consistency with the specialist information. The meetings were also an opportunity for students to get together as a team and discuss and plan their work.

The assessment was based around three staged written reports, (i) a proposal, (ii) an interim report and (iii) a final report. There were two presentations, one at an interim stage and the final at the end at which industry could attend and provide feedback to the faculty. Peer assessment was included with 100 points allocated to be assigned across the group. To help in providing peer feedback early enough so that approaches could be changed there was a preliminary peer assessment at Week 7. If the peer assessment was showing potential problems faculty then could intervene. Reports consisted of a main group report on the design with appendices for each sub-discipline given above. There was also a request for a copy of a typical set of meeting minutes. Written feedback was provided for submitted reports. There was also an opportunity for face to face questioning on the final report with the group. In this questioning session the main part of the report and the specialist sections in the appendices could examined.

For the faculty the challenge each year is to find a real-world project that is current that involves a range of challenges that requires involvement of most, if not all the six areas of specialization. The 2019 project with the coastline path and cycle-way extension was seen as almost ideal as it involved all key areas. The involvement with the industrial companies was seen as a great benefit, especially when they provided senior and more junior personnel to be involved who could provide information to the students including explaining their typical working day. The support was seen to raise the profile of the company too, especially with the students who would soon be graduating and entering the workforce. Besides the main company undertaking the project there was also involvement from engineers from other companies working on the project.

The real-world problem provided relevance and an authentic project for the students. Being locally sited allowed the project site to be visited as well as an appreciation of the significance to the community. Students gained experience at seeing the site. Comprehensive data from surveys and local consultation gave significant material for consideration. Students within the group each took a specific technical role so each could focus on their specific area and apply their own technical knowledge to the problem. Weekly meetings allowed discussion within groups and with the instructors who were the technical experts. Regular reports ensured the development of the design.

Further details on this capstone project course can be found at [45]

6.7 Non-capstone Projects

Project work was been traditionally encountered only in the final year stage of an undergraduate engineering degree. However, there has been a movement towards introducing project work at earlier stages in engineering programs, including at the first year level. What are the benefits of adding a project course earlier in a program, as these courses will need to either replace another course or become an additional course? The benefits could include introducing more design experience and knowledge, making connections and consolidation across a range of technical subjects, experience of project management, developing teamwork, problem-solving and communication skills, and even enthusing students by directly involving them in engineering project work. There is an added advantage that earlier exposure to project work can help improve the capstone experience. For example, learning to work more effectively in a team, or with project management can help make subsequent projects move forward more effectively. Beyond scaffolding for the capstone, the use of project work to help students integrate knowledge together from different courses can be valuable. The compartmentalization of subjects can reduce the understanding of the interconnections between the material taught and projects can break that barrier. I have seen this with the already mentioned third-year project that uses microcontrollers and brings together sensors, analogue and digital electronics and coding. When put together with a team project problem to solve with a design and build, the mix becomes an interesting learning experience for students.

Non-capstone project work in recent years has been included as part of a problem-based learning (PBL) approach (discussed in chapter 5). Even if a PBL approach is not being explicitly followed the motivation for use will have similarities for the reasons for PBL.

The level that the project is introduced can have an affect on the objectives and features of the project. The name capstone is used for the final year project of a program to indicate that the notion that the project will draw from many aspects of the courses in the program and it is an experience that will have synthesis of knowledge. Of course a first year student cannot be expected to undertake a capstone project as they have not yet acquired the knowledge that they will accumulate over the subsequent years in the program. So, design of projects at earlier program stages require careful planning to provide some level of success by a competent student. Adding to the challenge is a possible time constraint, as a full academic year project may not be possible. It may need to a one-term project or less if it is part of a course. However, as mentioned earlier the pedagogic advantages have been seen to be valuable enough that project courses are implemented into the curriculum.

In a moment we will look at some examples in the published literature. First there needs to be some consideration of the different levels and some of the constraints that come with those levels.

First year. This stage is traditionally where fundamentals of engineering and science are introduced. Students come from different schools and colleges, different parts of the world and possibly making a return to education, having done other things after leaving school. The prior educational experience of students may be quite diverse and this is why the first year is often used as a way to equalize the cohort's knowledge to fill in gaps and move people forward for the material that is to come in the second year and beyond. As well, many engineering programs have a general entry and so at the end of the first year there will be a need to decide what discipline of engineering the student will progress to. This often requires exposure to the various forms of engineering to help the student and program coordinators assess their suitability for the various options. The first year also has perhaps the largest cohort of students across the various years, assuming steady intake at first year and some level of students leaving the program or failing to achieve the academic level for progression. That number, if large, can cause logistical problems with resources, such as project space in laboratories and workshops, instructors and TAs. Capstone projects will still need to be resourced and will likely take a priority for space and equipment. Any project at first year (or any other year) will need to have some assessment on the availability of resources. Despite these challenges a project at this stage can lay foundations for better understanding design, teamwork and the other important aspects that project work deals with. It can also help the student see what area of engineering interests and suits them, if they have not yet decided. The project can enthuse and show the students the more practical aspects of engineering, unlike many classroom-based courses.

Second year. This year follows the first year that provided that fundamental and baseline knowledge. Now, more discipline-based courses are taken and sometimes those courses are in a sequence of stepped knowledge courses that form core knowledge for the chosen discipline and program. That stepped knowledge usually is continued in the third year. If not already encountered this could be the year that communications-based courses are taken along with non-engineering electives, in the arts and social sciences, to help broaden the knowledge of the student. The program in this year is again often full. Squeezing in projects into the timetable can be a challenge as something needs to be omitted or moved. At this stage though the students have more of a solid foundation. The advantage of a project course here is it can act as a way of drawing together the various other courses which may appear to be in distinct blocks of separate knowledge, such as analogue electronics, digital electronics, computer programming and advanced mathematics. The end of the second year places the student at the midway point of their programs and so the inclusion of a project can help students overcome any potential mid-program 'blues' or tiredness with studies. The self direction of studying as well as a focused purpose of a project can help students feel there is direct progress in their studies. This may also be the year before undertaking coop placements and having project work experience may help with securing and undertaking a work placement.

Third year. This is the penultimate year for the student's program. The courses become more detailed in their discipline and prepare the students for their more specialist courses in their final year. The students may be returning from coop experience and be keen to undertake project work. A project at this stage can help prepare for the capstone project experience as well as draw together technical details from across the range of topics from the previous two years and whatever is learned in the third year. More information on design and design processes can prepare the students for their capstone design course. Because of these reasons it is this year that can be a valuable year to include a project course. Again, there is the challenge of fitting a project in and potentially replacing another course in the program. However, the advantage of the synthesis work of projects and a chance to help reinforce concepts and learning for the final year make it attractive to make the effort to somehow include a design project.

The role of projects in a program can be useful and used in a variety of ways. The form of project work makes the learning more student centred. The student undertakes the work and applies their knowledge in the process and gains new knowledge as the task and solution is pursued. The knowledge applied can break beyond the boundaries of a single course, so there is potential for relating the knowledge from a range of courses to solve the problems within a project. This could be considered as the *relational* stage in the SOLO taxonomy. There is also the benefit of drawing students deeper into engineering, which can both enthuse the student and help establish their current position in the broader engineering community.

There have been some interesting and creative attempts to integrate non-capstone projects into the engineering curriculum. In the first year , when a project can enthuse and challenge a new engineering student, there has been a number of reports on various implementations of projects. One interdisciplinary project had 40 students from mechanical and 20 electrical engineering programs coming together into project groups of 5–7 students [20]. This project operated for 15 weeks and incorporated technical work in a robotics area, along with aspects such as teamwork, creativity, leadership. Besides professors from the two departments there was added support from final year engineering students who becomes the leader of the team. This student received training in leadership [20]. The assessment was based around deliverables, including a planning report (due in week 3), a final report (week 13) and oral presentations and defence. Rubrics for these assessments can be found in the paper [20]. Evaluations from students were done over a number of years and interestingly the very first year of the project running there was a low 3.9/10 overall evaluation from the mechanical engineering students, whereas the electrical students responded with 7.3. The reasons given for the low response from the mechanical students included:

- unfamiliarity of doing interdisciplinary projects by the mechanical engineering faculty, whereas the electrical engineering faculty had some experience;
- some project activities occurred towards the end of the term, at the same time the report was being written, so students felt overloaded;
- content in the project was also covered in other areas so again an overload of work was felt by the students;
- work was unevenly split by students;
- the team leader, a final year student, had not experienced the project as a first year student as this was the first year the project had run.

The following year modifications and improvements were made and the overall score rose to 6.1/10. The authors also highlight some of the improvements and they include: improved communication and coordination amongst professors; paying attention to avoiding overloading of work; delaying use of a student leader until a student who had done the first year project was available in the final year; grading of course work should only be done in the context of the project. One interesting features to note used within this project was the use of a Belbin test [16] with the students to make the project groups more mixed. This test is an inventory test that can categorize the individual taking the test into one of nine roles. These roles are: resource investigator, teamworker, co-ordinator, plant, monitor evaluator, specialist, shaper, implementer and completer finisher. Descriptions of these different roles can be found on the Belbin website [16], which includes details of some strengths and weaknesses. It is worth looking at these descriptions to see the variety of team member roles that have been categorized by the work of Dr Meredith Belbin. Mixing students with different roles can ensure a diversity of traits and perhaps a coverage of intrinsic abilities that can make a project more successful.

Another reported first year project course was done with chemical engineering students, who had to build a hydraulic jack [227]. This study reports on the integration of the project across two courses, one on the principles of chemical engineering and the other chemical engineering fluid mechanics. An aim in this project was to connect and integrate material from the two separate courses, with the effect of developing the student's critical thinking skills. Over 200 students undertook the project in teams of 4 to 6 members. In planning the project the instructors designed the course to be in stages with sections affecting the cognitive, pyschomotor and affective domains. As well the formative and summative assessments we distributed across the stages. For example, a problem restatement and identification stage could use the cognitive domain and included formative assessment. Whereas a demonstration stage could involve all three domains and have summative assessment within it. Like the previous example study, the students had their traits assessed, but this time they used a DOPE personality test. In this test individuals are assessed and are grouped into four categories-based named after birds, dove, owl, peacock and eagle, hence the acronym name. In general, the investigators found a balanced distribution of the personalities across the teams. The project was assigned in week 5 of the semester and started the following week. At the end a demonstration of the hydraulic pump (that was specified to lift a 500 g load). Team members evaluated each other were also given the chance to anonymously give opinions. Reflections were also undertaken to self-examine the project. Summative assessment was done in the demonstration, project report and peer assessment. Formative assessment was done in the problem identification stage, project consultation, self-reflection and an exit survey. This final survey was 15 questions that looked at how the project had helped the student in various aspects as well as their opinions on the project experience. The authors note that "90% of students agree that the project managed to enhance their understanding of chemical engineering fundamentals" [227]. There were acknowledged agreement that their skills in communications, time management, team player and leadership had been helped by the project.

As mentioned, I have been the instructor for a third year design project for electrical engineers for a number of years. This project is a one term course and so requires the students to be guided through the 13 week process carefully. Groups of 5–7 students are randomly selected into teams. The projects are decided by the students, with suggestions provided, and sometimes there are clients with projects. One year there were three artists who provided projects. To help with the group organization a formation form is provided that requires students to complete, which includes questions I have mentioned earlier, Table 6.2. This requires students to consider how they will communicate, make decisions and respect roles. This could be considered as a professional behaviour guide. There is also a place for students to consider what a good team member is like. As the class size can be over 100 there is a likely chance that students have not worked with some, or even all, of their team members before. The completion of this form along with the consideration of the project becomes a focus of teamwork and allows students to work productively from the start. There are construction exercises too with the soldering and testing of a kit microcontroller board. This also is another focus for the students to interact and start to act as a team. Mistakes in assembly and software testing cause the team to work together to solve early problems. The students can be moving through the Kolb learning cycle when problems can occur (a concrete experience), considering where the problem may lie in either hardware or software (moving through reflective observation to concrete abstraction), and then considering solutions and steps to resolve the problem (active experimentation). This helps prepare them as a team with the other challenges they will encounter during the project.

The requirement for the students to decide upon their own project causes them to consider what they are capable of doing. There is an implicit reflective process here of having to assess what they are capable of from their prior years of experience. Sometimes experienced students can act as a guide to this process and leaders can occur. The free format can be intimidating to groups, to counter this two instructors and teaching assistants can help guide the groups in formulating their ideas, as well as provide suggestions.

After selection of the project there are the decisions in the implementation approaches. This requires students to consider the solution and then the components they need to purchase. Using data sheets for components they can select and order the components that they need, through a teaching assistant. This may be the first time that they are selecting components and have to become familiar with how to search for appropriate components. This is different from a conventional laboratory where components may be specified. Each team has a budget. This makes students aware of budgetary constraints, trade-offs that may be needed as well as considering lead times for components to arrive. From a situated learning perspective there are a range of elements in the project that can make these third year students learn aspects of practical engineering work and make the students be aware that they are undertaking the engineering process and applying their knowledge and acquiring relevant experience.

The assessment work for this course has been outlined in Table 6.4. There is also the commencement of regular reflections that continue throughout the project. These reflections can be on something of their own choosing, connected to the project. However, prompts are given that link to the current stage of the project. For example, an early reflection suggestion is the consideration of changes to how the group formed and got underway.

Along with the scheduled project laboratory work there are also scheduled classroom session. These cover technical areas like microcontroller programming and common sensors, as well as working in teams, design aspects, life cycle analysis, sustainability and as the university is in Canada there is inclusion of an Indigenous learning bundle on Indigenous environmental relations [252]. Beside the instructor led sessions there are regular opportunities for students to give presentations, such as early sessions to reveal what there project will be and the work breakdown and explaining their designs. To later sessions include one significant thing the group have learned from the project and final presentations. These grouped sessions provide an opportunity for all to see what other groups are doing and how they are approaching their projects and what they are achieving. With about 20 groups these sessions do take time but provide a broader perspective than just what is going on within their own team and project. The final laboratory session becomes a final showcase with TAs and invited faculty being able to see the projects demonstrated at the lab benches and talk to the students about their designs and builds.

From a situated learning perspective there can be seen to be some major steps in this electronics project. The process of moving from an idea on paper, created from discussion within the group. Through the allocation of work and its scheduling. Having to make design decisions and considering alternative approaches. Considering aspects of safety, user requirements and application. The learning of skills, such as soldering, using test equipment and combining hardware and software together, with the associated debugging. These all combine together to make a authentic situation of engineering. The sense of achievement at the end is often discernible in students.

Some students bring experience to the project, perhaps from co-op or their own outside the curriculum interests. The opportunity for peer learning between students is strong and that brings added value, to the person giving the knowledge and the person receiving. Students also see how others approach problems and how they work. Early on in the sessions as project ideas are forming, the instructors and TAs need to be active in discussion with groups their ideas and what is and is likely not possible given the constraints of the time and money resources.

The situated learning perspective for these non-capstone will be similar to that of the capstone project, see Figure 6.1. There can be some limitations on the expectations in the situated environment, as they are not final year students. However, undertaking projects and doing design and construction work will help develop the identity of being an active participant in engineering activities. Any lack of experience and understanding of mediating languages can be overcome with the guidance of others who do understand them. This could be instructors, teaching assistants, technicians, or even peers. New equipment may be used too. This again requires further learning and helps the students to become more familiar with specific equipment and the process of learning new equipment. The roles students undertake brings responsibilities and different perspectives. Teams of students will plan and work on different elements of a project. This is not like the student doing the same problem in a class assignment. The individual role provides a more personal task and ensures each has to find their own solution, using others to help with determining solutions. Encouraging reflection on how solutions can be created and formed can help generalize the learning and help in the future. Of course, not everything goes well and so any poor teamwork or dysfunction in groups can lead to weak performance in the project and even a bad experience. Instructors here need to pay attention and carefully balance the individual and team response to the challenge as well as provide ways to safely learn and make mistakes that do not jeopardize the cumulative learning in the project.

6.8 Chapter Summary

In this chapter we have looked at perhaps the most significant and important areas of experiential learning in an engineering program. Often the only formal project work is down within a capstone project or design course. In this chapter we have looked at these courses and their development and implementation in undergraduate engineering education. However, the capstone stage is not the only area where projects are being introduced in programs. The engagement and learning where projects bring, mean that they are also being introduced at different stages, even in the first year. The decennial capstone design surveys have shown that there is a move towards team-based projects and multidisciplinary projects too. To understand the learning we again have applied a situated learning framework and this allows us to see the elements that are generally encountered. For any project that an instructor is involved with it may be valuable to think about the situated learning in that particular environment. Consideration of assessment approaches is valuable, with opportunities to move beyond a reliance on just a final report. Some models on assessment show that it is possible to try and assess the personal development, teamwork, design approaches as well as the solution that is presented.

7

Cooperative Education

The inclusion of a period of time within a degree program for students to go and study in the workplace, relevant to their degree, is long established. This adds time onto the degree program but both students, university administrators and perspective employers see the value. What is interesting is that this is one of the few periods where a university instructor is not directly involved in the learning experience. There maybe a departmental coordinator and instructors may be assigned to visit students, but day to day supervision and guidance is passed onto the employer. The companies and institutions that employ the coop students are the hopeful destination for many newly qualified graduates and their supervision and training experience is trusted.

In this chapter we will look at this form of experiential learning, the role of co-op and industrial placements in engineering programs and the benefits of this experiential learning. For engineering students deep in their academic learning a period of stepping outside the university campus and having a period of work experience can provide opportunities in a variety of ways. They can directly see the application of their academic learning. Most will encounter the limits of their knowledge and learn more. Some of what they learn can be very specific to their area of employment but valuable nevertheless for the experience of learning the specialist knowledge, as well as for potential future employment with the same employer or those in a similar field. Problem solving will be regular and they will see licensed engineers and other professionals working to resolve those problems. That opportunity to work with teams of engineers and other professionals will provide valuable experience for the student. The employer benefits by having students fresh from classes, they participate in the education of the students directly and they have the opportunity to find future employees.

Perhaps the student is not in a formal co-op program, then it is possible for the student to still gain experience through summer employment in a related industry. This is a similar to co-op but is perhaps shorter in duration and consequently a shorter potential project. There is no supporting infrastructure of a co-op office or departmental advisor.

DOI: 10.1201/9781003007159-7

7.1 Origins of Cooperative Education

Although there had been apprenticeships for some time, the cooperative learning in the US started with the field of engineering at the University of Cincinnati in the school year 1906–1907 [206]. The founder of the idea was Prof. Herman Schneider. The details of the creation and initial operation of the cooperative program at Cincinnati was recorded by W. C. Park who taught in the Department of English at that university in 1916 [206]. The record provides interesting insight into the then new approach to engineering education, including operational details on the implementation.

The problem that Schneider was trying to solve was to provide a suitable education for the engineering students, but realizing to have laboratories that included all the costly equipment that was within the industrial plants of the times was financially impossible. Parks mentions that the existing program and industrial experience that engineering students went through before the cooperative program was created was four years of study and then two years of an apprenticeship in industry. The solution to the costly lab equipment and experience problem that Schneider came up with was to intersperse the class, and more theoretical experience, with industrial experience. So with a period in the classroom and then in industry the students could gain the required knowledge from the university as well as working on the equipment currently being used in industry. Now the students were not only getting to use their knowledge, working with processes and equipment that was needed, but they were also working in the settings they would be seeking employment. Further, the students were being employed and earning money. This had a benefit that the pool of possible students could be broadened as the cost of being a student could be offset by the earnings from the work.

From a situated learning process there is an integrated and arguably an accelerated learning approach occurring. There is a combination of scholarly study mixed with the real world understanding of how the textbook knowledge can be applied. Content and context are combined. Students work with professors in university and practitioners in industry, learning from both.

Park gives the definition of cooperative education in 1916 to be:

> "... the coordination of theoretical and practical training in a progressive educational program. Since the agency which furnishes the practical experience is always some branch of actual industry, the reciprocal relation between school and shop permits the fullest possible utilization, for educational purposes, of equipment used in commercial production. Obviously, the arrangement of alternating periods is a mere administrative detail. From the employer's point of view, the most important elements of the cooperative plan are: First, the selection of workers; and second, the awakening of an enlightened interest in their work through coordinated instruction."

"From the standpoint of the school and the student, the most important feature of cooperative education is the realization of theory through its practical applications. In a very literal sense the studies in the curriculum become "applied subjects." In the use of of the word "cooperative," emphasis is placed not only on the kind of training, but also on the relation between school and industry, and on the method of bringing them together." [206]

Park highlights the situated learning nature of this form of experiential learning. There is recognition of the equipment used to "the fullest possible utilization", along with the '*workers*' (note not *students*) had their "awakening of enlightened interest in their work". In the second quoted paragraph he notes and emphasizes the combination of the academic with industry to the education.

The early cooperative education scheduling was different to the blocks that students encounter today. As the supporting industries were often local to the university the students worked with a weekly cycle initially in the cooperative program's development, but this soon changed to a two-week cycle. Clear details are given in Park's book of the shop work records that were kept. These indicated the date and duration (including hours off) and the department and nature of work. Weekly work hours were 49.5 hours. Records also show the detail in recording the work in offices, planning departments and machine shop for example (in the mechanical engineering degree). There is also note of unplanned repair work in exceptional circumstances, such as the Ohio floods of 1913. There was staging of the experience over the years and Parks records for electrical engineering students:

"The first year's work was in the foundry; the next year and a half in the machine shop; the next two years in the commutator, controller, winding, erecting, and testing departments; and the remaining time in the drafting rooms."[206]

The support from industry for the cooperative education is noted in the book, from the early ideas of the program, which helped get it going to the continuous support. In a paper at the Society for the Promotion of Engineering Education in 1907, just after the first year of the cooperative program, Charles Gingrich of Cincinnati Milling Machine Co. gave a supportive paper which included:

"I do not mean to imply that our shops are full of secrets, but I do want to emphasize the fact that they contain a vast number of things to be learned; that the only place to learn them is in the shops; and that the best way to do it is to start young and take plenty of time. The chief criticisms of modern technical education result from the fact that we try to take shop into the school, whereas we should bring the school into the shop. The cooperative plan does bring the school into the shop." [206]

From this early and successful start at the University of Cincinnati the cooperative education program grew. Other universities in the US started to follow and include cooperative education, including Northeastern University, University of Detroit Mercy, Georgia Institute of Technology, Rochester Institute of Technology and the University of Akron. In 1919 the GM Institute set up a 100% mandatory cooperative scheme [59]. This institution later became Kettering University. From these pioneering institutions the cooperative programs grew across the US and became an established part of many, if not most, undergraduate engineering programs. There would be similar developments in other countries. In 1957 the cooperative education was started at a new institution in Ontario, Canada, at what is now known as the University of Waterloo [178]. Since that time the Waterloo has become to have one of the largest cooperative programs in the world.

When comparing the histories detailed about the birth of the cooperative system and the University of Cincinnati in 1910s [206] and the University of Waterloo [178] in the 1950s both found that the people within established academic circles were less than positive with the change. Whereas, it was those in industry – the engineers that Schneider consulted with his initial idea of cooperative education and the businessmen that wanted to found a new institution in the Kitchener–Waterloo area – who were the most enthusiastic and who saw the value to the education of students. In the case of Waterloo the resistance came from "delegations from nearby institutions who came not to praise them and find out more about what they were doing, but to criticize their efforts" [178]. This contrasting view of the merit of cooperative education may have come from a lack of understanding of the benefits and value to students and their employers, as well as the perceived weaknesses and deficiencies of the students when they enter employment. Let us now look at the benefits to both students and employers in the next section of this chapter.

7.2 Benefits of Cooperative Programs

At it most fundamental cooperative education is the full-time employment of a undergraduate student into a position that relates to the students degree program. The time at the employer will mean extra time to complete the degree, often an extra year. The employer for an engineering student may be in the industrial sector, including manufacturing or research and development. Or the employer could be in some level of government or public institution. This employment will have been approved by the student's home academic institution who will usually give an appropriate designation on the student's degree certificate, upon successful completion of the academic and cooperative requirements. However, the cooperative education is much more than that providing benefits to both the student and the employer. We will consider these in turn and then in the next section we will focus more specifically on the learning.

7.2.1 Benefits to students

The benefit to the student is to have an opportunity to gain appropriate employment. To get the co-op employment it is typically a competitive process. This is just like trying to get a position after graduation. So students will have to learn and develop skills in resumé or CV writing, as well as interview techniques. Often the cooperative office, or similar, at the home academic institution will provide help and guidance. There may need to be a preparatory course that has to be taken, during the normal academic program, to prepare a student for their first co-op position. Even the application and selection process for co-op work positions can provide a valuable learning experience for the student in later years.

When successful the student can enter employment for a number of months, commonly a year. Possibly this could increase to a possible maximum of 18 months, with employment occurring during the usual academic summer breaks. One obvious benefit to the employment is the salary and a chance to earn money for living expenses, savings and costs to being a student (such as course fees). Employers will be expected to offer appropriate salaries.

The employment may be the first time for many students that they are employed in an area related to their degree program. They then have the opportunity to see first hand how professionals operate and how work in their intended professional area is conducted. Their work will be overseen and often they will be assigned to a manager, team leader or appropriate engineer who will guide them and their work. Specific areas of employment can be diverse, even within a discipline. They may be in a manufacturing environment and could be working on production, testing, quality assurance or development. In government or public institutions they could be working on research, development and public works activities. Here comes another benefit, in that a student can see whether a particular area or type of work interests them. For example, some students may not consider quality assurance (QA) work in their engineering degree, but experience of seeing this type of work may make them interested in testing, calibration and the data analysis related to QA. Or, they may find that an area they had initial thought they wanted to work in was now not as interesting to them, as they anticipated. So it can help prevent a poorly informed decision at an early career stage.

Within the employment the students will use their academic knowledge. There then becomes a stage of applying and seeing applied what they have learned in the classroom or laboratories. They may work on specific equipment or tools that they have not have encountered at university - the motivation Hermann Schneider had, of the need to expose students to leading and costly industrial equipment [206], is still valid. They can learn from the various experience of the tasks they are assigned. Either from directly working on a problem on their own, or through working with their supervisor or as a team on a specific problem. The work areas will be authentic and real problems to be solved, though some employers may have a training period for co-op students. The work itself will often be in a specialist area and will develop specialist knowledge. To do the work the student will draw on their prior experience, which will be mostly their classroom education in the courses in their program. The work they do will then give another set of experience such that when they return to the campus for their final stage of education they will have a context for some of the material they meet in their courses. As a personal example, in my own co-op placement I worked in an company manufacturing optical fibre. In particular I was using a computer to control test equipment and record measurements. When I returned to my academic studies and did optoelectronics courses I was familiar with the why and how to measure the losses in optical fibre at different wavelengths of light, as I had used and developed a test bench to do that. As well, the computer control of technical equipment and the statistical handling of measurements fitted with other courses. This experience was unique to me. But my fellow students had worked in other areas and they had their own experience and connection points to course material. Sometimes the overlap between what was worked on in the work term and what is taught on return is strong and other times it is weaker. The benefit overall of seeing aspects of what was done in co-op occurring in their technical classes can have impact in validating the co-op work as well as giving perspective and appreciation of the knowledge learned in the work term. This is shown in Figure 7.1. In my personal example I had some optics, electronics, measurement and computing knowledge before going into the work placement. These areas all came together in my co-op placement. On my return to the campus and classes I found courses on optical fibre were supported with first hand knowledge of handling fibre and doing tests that produced the type of graphs that were shown in the classes and in books.

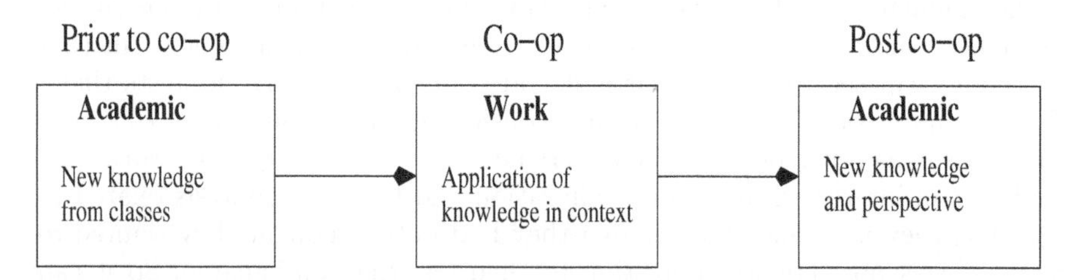

FIGURE 7.1
A representation of knowledge from the academic setting, into co-op work setting and then back into the academic setting after co-op.

Beyond the technical work there will be learning of other process-based skills such as interpersonal communications and report writing. This can lead to a better understanding of the importance of these skills and an appreciation of the need to develop them. The different approaches to learning will help and may lead to a better understanding of leadership and management. Combined with the interpersonal skills will be the development of a personal network, established during the co-op. This may include managers, team leaders, work colleagues as well as fellow students who were at the same company on placement. The fellow students maybe at the same university as the student or at another institution. One or more of the colleagues who were permanent employees at the work place, could also become mentors to the student. This could be only over the duration of the coop placement, but it could be beyond. Mentors can provide career and psychosocial help [76]. Although the majority of help provided and sought will be of a career advice nature, there may also be the need for more personal help and advice. In a relatively brief coop placement the time for the student to develop the trust to establish a mentor can be short. However, mentors maybe proactive and reach out to provide the guidance that a student may need. At the very least the supervisor should give some level of guidance and mentoring.

Towards the end of the work term there will usually be a need to write a report for the university, to be assessed by the cooperative education office and/or the home department. This will give more report writing experience on their personal experience and learning. As well, this is an opportunity to reflect on the work done and what has been learned. More will be discussed on reflection in the next section.

Following the work term the benefit of the specialist knowledge and the experience provides an area that may make the student attractive to other employers, or the student may have attracted the attention of the coop employer who may be interested in considering the student as a future employee on graduation. The employment can start to distinguish the student as they now have something different to record on their resumé, as much of the experience at a university program will be similar to their cohort. Citing a co-op placement on a CV or resumé and the experience obtained there can help to separate that student from others who did the same courses and laboratory work.

Since co-op learning can be a key difference between students studying effectively the same academic program, but with some doing the co-op option or not, the question can be raised; is there an advantage from doing coop when gaining post-graduation employment? One study by Walters and Zarifa in 2008 used Canadian data, collected by Statistics Canada (the Canadian Federal Government's national statistics office) to examine if the was benefit to college and university students doing co-op in terms of employability and earnings after graduation [276]. Because of the data they accessed their study could also examine the difference due to gender. They reported finding:

> "In terms of securing full-time employment, coop programmes provide the greatest advantages to male college graduates and female university graduates. With respect to earnings, coop programmes provide the strongest advantage at the university level, particularly among males." [276]

This is a snapshot of data in time (from a Statistics Canada survey done in 2000), in one particular country. The situation may be different in other economic situations and in different countries and regions. The authors do note that there is a 'sizable wage gap', between those graduates who have taken co-op programs and those who have not [276]. This is even after accounting for sociodemographic factors. They surmise from the analysis that this apparent coop advantage is due to the value of the credential, and not related to any difference in the graduate characteristics from different programs [276]. They indicate that the perceived value of doing co-op seems to count with employers.

Schuurman and colleagues, at the Pennsylvania State University, also undertook a study which was published in the same year as the previous Canadian study [234]. They had some similar findings. Their study examined exit data of graduates from the College of Engineering and looked for the impact of work experience on post-graduation starting salary, cumulative grade-point average (CGPA) on graduation and receiving a employment offer before graduation. They also considered if these impacts were different between male and female students. They found that there was an increase in starting salary for those with work experience, and job offers before graduation were more likely. As for the impact on the CGPA, that was found to be "only marginal" [234]. The also found that the impact was equally beneficial across genders.

7.2.2 Benefits to employers

What about the benefits of cooperative education to employers, as the majority are not primarily focused on education as their business (accepting that some co-op students may find work in universities)? As realised by the forward thinking business people who helped established the University of Waterloo [178], there is a recognition that participating in the education of their future employees or the employees in their business community may be a valuable and useful thing to do. Having students see and work first hand in their industry and business for a short time during their education can be seen to have distinct value. That can come in a variety of ways, with students understanding better how to work and contribute in business or public service work. Seeing how academic knowledge can be applied and what parts of their academic experience needs to be developed, can help enhance the learning and development of a student who may soon become a new employee. Having a student who has had a year's cooperative education experience can produce an employee who is quickly contributing. Whereas a new graduate without cooperative experience may need another year of employment to get to a similar level of experience. The expectations and commitment to a new permanent hire being more significant than a six- or twelve-month placement.

Friel examine the benefits of employers by surveying cooperative education directors in companies, mostly in Texas [100]. Of the 691 surveyed 18% responded. The findings showed strong support for co-op. Some reported benefits were cost reductions in training, the reported abilities of new hires who had co-op experiences and also access to minority groups [100]. Despite the training cost reductions there was indication that hires with co-op experience had higher starting salaries. This matches findings mentioned earlier in the benefits to students. Friel's survey also looked into potential problems with co-op and the responses included cultural barriers (such as not knowing business methods or how to dress), continuity of work, lack of exposure to the technology in the work, not showing interest in the work, cost, what to do with co-op and no minorities [100]. It should be noted that the large responses were either no answer or no problem.

The hiring of co-op students is also a way of having an extended recruitment process, which is another key advantage. The hiring stage of a co-op student is one level of selection, but to be able to have a manager and engineers assess the student in the work environment, over a number of months, can prove to be an effective and efficient approach to hiring new employees. If the student excels in the work environment, and there is are positions available, then there can be moves towards a permanent hiring. The fit between the student/employee and the employer has already been tested.

As the employee colleagues of the co-op student help to educate the student, there is also the benefit of the student helping his employer and colleagues by bringing in skills. I learned this first hand when I had my own previously mentioned co-op placement in the 1980s. From courses I was proficient with computer programming and interfacing computers to test equipment, along with experiment design. This ability, coupled with being an extra resource as a student, allowed me to be valuable on a project to further automate test benches for measuring some of the optical properties of the fibre. This detailed work was ideal for a student under regular supervision by an engineer. It also did not tie up the engineer with the time consuming process of code development. That work extended into being able to help select and order new equipment for further tests. This was my first hand experience of the symbiotic benefit between employer and employee. I could develop a project much faster than an engineer who had it as a 'corner of the desk' project, amongst a number of other projects. That experience directly led me to further employment and postgraduate study. Later, as I worked as a manager, I knew and understood the benefit of a co-op student and hired them to work in my teams. Although a single personal experience this shows how the benefit of cooperative education can perpetuate. Many co-op students may go on to become engineers, team leaders and managers who can make the request for taking on a co-op student and understand the type of work and experience a student needs. This can be taken a stage further as a hiring manager can pass a co-op student onto a team member who they want to develop as a team leader. Having the engineer supervise a student is a way to help that development and assess the further training that team member may need. So, co-op students can also be an integral part of management development too.

7.3 Situative Learning Perspective of Cooperative Education

Applying a situative learning framework to cooperative work terms can allow broad insight into the student's learning during co-op placement, as it is both different and linked to the traditional on campus learning [81, 82]. The interaction the student has with different people encountered in the workplace, the projects worked on and tasks involved with the placement, as well as the equipment, tools, technical language and other mediating objects in the workplace will all be an individual learning experience for a student. There will be common features and stages for coop students and we will look at those here through a situated learning perspective. A simple graphical summary of the situative learning model for coop is given in Figure 7.2. This can show the change from a student applying for a positions to being able to work independently. Another commonly used model for the learning in co-op is the reflection process [117, 118, 164, 216]. We will make some reference to this model too, as it can complement the situative model.

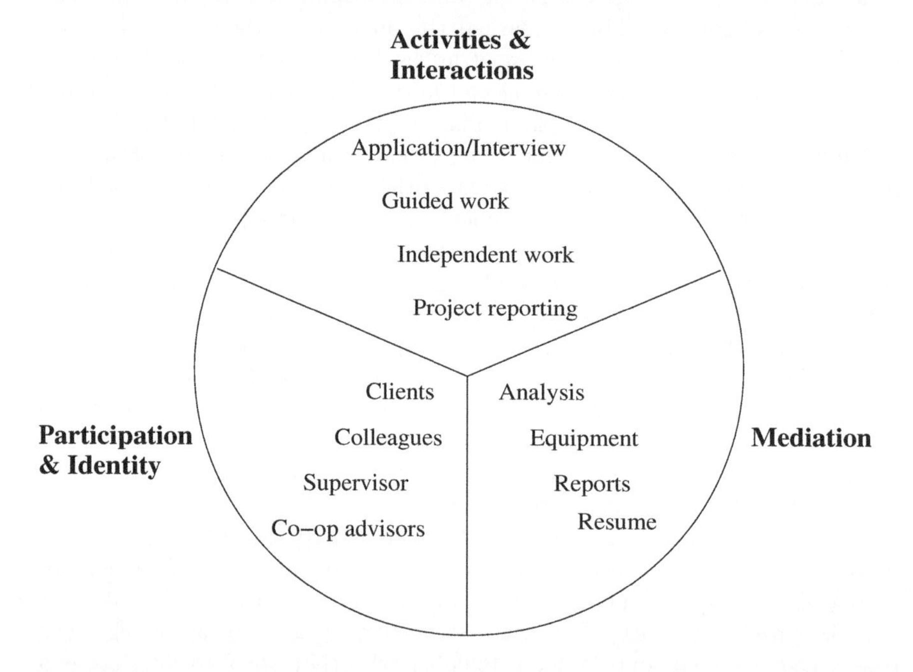

FIGURE 7.2
A situative model of cooperative learning focused on three areas of interaction, mediation and identity. The stages, from a student applicant to operating relatively independently in the work environment, are shown by the stages of moving from the outer ring of the circle towards the centre.

The work term learning will start before leaving the campus. The preparation and support for the student will be provided through some combination of the cooperative office and academic and departmental advisors. This is denoted in Figure 7.2. The preparatory support will be mostly focused on helping the student secure appropriate employment. Learning modules and courses are often provided by cooperative education offices. This will help the student in the preparation of their resumé or *curriculum vitae* (CV). In building this document the student will be required to reflect on what is important to highlight to a perspective employer. This in itself can help the student, as they consider their academic education to date and the key stages within it that an employer may find of interest. Along with the formal education there will be the employment and other relevant activities outside of the academic environment. This holistic examination looking at their collective experience, knowledge and learning will be helpful to the student to assess their strengths and weaknesses. For the perspective employer the resumé or CV is used as a filtering process as well as a starting point for possible future discussions in interviews. Co-op advisors can help the student learn how to better shape their resumé and this can help in the future as that document grows with experience. This may be the first time a student has sought employment, particularly in a professional area, and this advice and guidance can help the student in a significant way.

Logically following from the resumé building is the seeking and applying for employment. Again the cooperative education office can help as there will be employers that regularly seek co-op students from their particular university. As well, that office will maintain details of companies that are looking to hire. Students may search on their own, perhaps even using the their own network through family or past employment. Here there is learning on how to move into professional employment and there will be interaction with others, from professional to individual supporters to help connect to perspective employers. The required self-organization, as well as the value of networks, can help with the student's learning of how to seek employment.

After securing interviews there will be the first encounters with representatives of potential employers. Interviews themselves can be intense learning experiences and the communication skills will become important here. However good the academic performance is, if it cannot be supported by competent communication skills then it is highly unlikely someone from Human Resources or a hiring manager will move the application on further. The interview process can become the first stage where the student applicant connects with the technical manager who is seeking the co-op employee. The manager will have looked through applications (perhaps many) and selected from those applications the candidates that appear to meet their requirements. With students progressing through the same programs, the academic details from an individual university program can look very similar. Besides the grade point average, the work experience and extra-curricula activities can become important differentiators and these are likely areas where hiring managers may focus in the interview. On the academic side the interviewing manager may focus on the academic interests and strengths, as well as any project or laboratory experience. The interview process will be a learning experience for the student. The communication of their abilities, as well as answering focused questions, can help not only the interviewers but can also help the student develop the more formal communication that they may be involved when working. If the interview is at the work location then there may be site tours and a chance for the students to first see the potential work environment. Fortunate applicants, who receive more than one offer of employment, may then have to choose the preferred co-op position. There could be a range of factors that come into this from the type of work, the technical field, the interaction with the hiring manager and members of the team, to factors like the location and pay. The process of choosing will require reflection on these factors and can be useful for the student to understand what is important to them.

The hiring process as the students interacting with campus advisors and employer's hirers. The process of preparing applications and success or failures with interviews gives the students opportunities to consider their qualities and attributes. Communication skills are tested and interview experience will help in the future. The identity of the student will start to change too. From being a student to being a hired employee in a professional environment. The process of visiting employers will allow projection of being able to work in such environments. Once successful in securing employment the student can further develop their identity as they commence engineering work and join a company or organization. The idea of being a co-op employee, with temporary work, and perhaps at the work location being identified as a 'co-op student' may still anchor the student's identity to being a 'student' but a transitioning can start as the student commences employment.

The work students do will likely be a carefully decided project and with a specific supervisor who may be an experienced engineer and possibly a team leader, or a manager. From early on the student will become aware of the hierarchy structure within the company, from the engineers, team leaders, managers, senior manager and perhaps directors and beyond. In the early stages of employment they will be seeing the hierarchy as well as the organizational structure. Whether consciously or not, they will be working out their position and relationship within the employer's organization. This may affect their self-identity within the company. As a student and co-op employee they may feel more of an outsider than if becoming a continuing employee. If a number of co-op students are employed there may be a tendency for the students to meet at break times and that can contribute to the way the student establishes their identity. Supervisors and managers should take their roles of supervising and educating their co-op employees seriously, so providing support and guidance. Depending on the size of the company or organization there may be formal training that the students can work on. However, the main learning will come with the work they are assigned to.

The work undertaken by the student is most likely guided by the assigned supervisor, but there will be a degree of autonomy. There will likely be a period of closely guided and assigned work. Perhaps to become familiar with conducting work on equipment or software tools. There will likely be scaffolding of tasks and training. Autonomy may come in the form of being assigned a task and being left to perform it after initial training and familiarization. The checking of the progress and performance should come at after some appropriate interval, such as the end of the working day, week or the end of the task. So, at the early stages of work there can be significant learning and adaption to the required expectations, with feedback being given on the progress. This process of doing and being assessed will be familiar to the student and may be a sought after and needed initially. Supervisors not providing the initial regular feedback may have a student that struggles with the degree of unsupervised autonomy, along with the complexity of the work. Early challenges can affect the self-confidence of the student and if the support is lacking, or the challenge is too demanding, this could even affect the morale (and prospective engineer's identity) of the student.

Mentoring may happen and be provided by someone in the co-op workplace. A mentor, who goes out of there way to help and guide the student, can provide career help as well as psychosocial help [76]. The career advice could relate to technical development and potential future career path. The psychosocial side could relate to adapting to the work environment and understanding the groupings and interpersonal aspects. This could be through watching a role model or being counselled by the mentor. The challenge a co-op student has is that their time at the workplace is of a relatively short duration and so is the time to develop a relationship of trust leading to mentoring.

The technical work, as mentioned, could involve, report reading for background knowledge, working with new equipment and in a laboratory, work site or office area. Software tools may also be used for design work, as well as analysis. Test results need to be examined and analyzed and there will likely be the reporting of findings. Here the technical expertise of the student could be drawn upon. There may need to be statistical analysis, calculations, use of engineering tools and coding. This technical work is the type of work the student is probably anticipating. It draws upon the majority of their courses and extends their knowledge in a real-world setting. This helps the develop the student's sense of identity of an engineer in training. Seeing the technical work of others too will place their abilities in perspective. The guidance of other engineers and supervisors will help focus the development of their technical ability.

The engineering student will also have clients, who in the work setting will require some sort of deliverable. The client may be their supervisor, it may be some other group within the employer's organization, or it may be a client external to the company. The checking of the submitted work will be an assessment of the quality of the work. It changes from being the assessment of submitted coursework, to the assessment of work that will use in a real-world setting. This feedback will then be valuable in setting of standards. There will be no rubric and grade assigned. However, there could be critical evaluation by the student's supervisor and fellow colleagues.

Besides the technical work there will be communications, such as oral and written reports. There will be regular group meetings, which will be ways to see the broader picture of the work the group is involved with. This will be important to set the context of the student's own work. This overview of activities and progress will help set the perspective of the student within the community. Knowing this range of activities can also open the opportunities for the student to request particular work assignments that interest them. An accommodating manager may allow these requests and this allows the student to guide their direction within the working community. As a student and at work for a temporary time, they may feel on the periphery of the working community. Allowing some element of choice, or other forms of self guidance, may allow the student to feel more empowered. Of course there may be constrains to the flexibility, as unlike at university, the main purpose of the employer is unlikely to be just education.

Some students may have a reluctance to take initiative. The change from a familiar academic environment to a working environment may be challenging for some. This may then require an adept supervisor to guide and perhaps mentor the student to go beyond the comfort zone. Academic ability may be a limiting factor, though many academic programs with a co-op option may have a qualifying level of attainment needed to progress onto a co-op placement. The application and interview process will also provide some level of filtering to provide a good match between the student and the position.

Towards the end of a project, which usually is towards the end of the term, a final report usually has to be written. This is useful for the employer, but it may also be a requirement for the academic institution and for academic credit of the co-op period. This document not only becomes a technical record of work done, but it can also be a focus of reflection on the work. Some co-operative offices may require an element of reflection in the report. The writing of the report will consolidate the thinking and understanding of the technical aspects and details of the work. This can help the student look at their technical knowledge development. The daily work on a specific project may have made them an expert on that particular aspect of the project work. This collection of the details in the report will help many students see their development. If the student considers and reflects on the learning done during the project there can be some benefit, whether that is with self-development, assessing their technical ability, interpersonal skills, or in their future career direction.

Harvey et al [117] and Lucas [164] have provided similar diagrams illustrating a conceptual model of looking at have reflection for a student involves a combination of the learning at university, the workplace experience and by the student themselves. Figure 7.3 is derived from those diagrams and shows

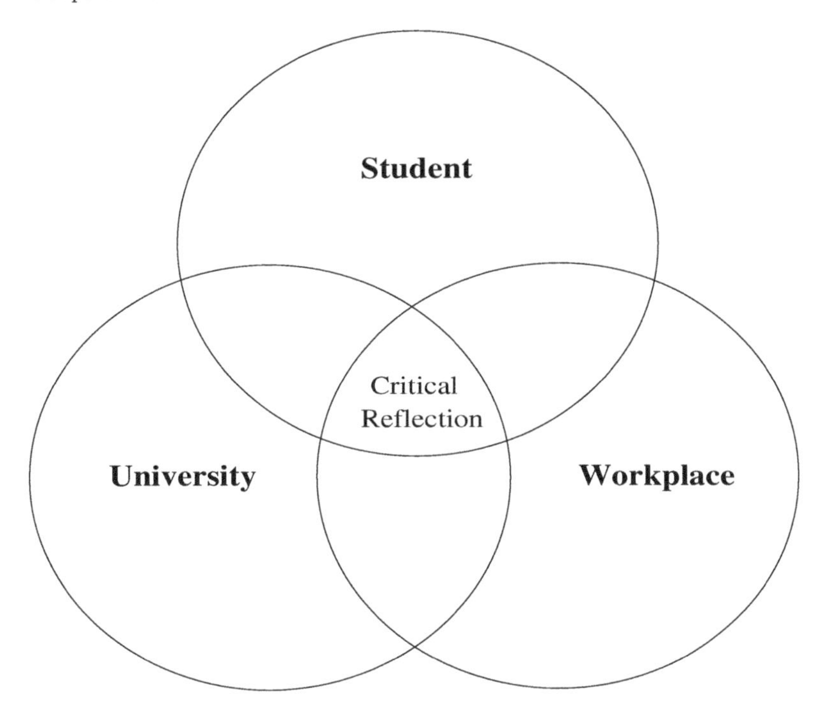

FIGURE 7.3
A conceptual model of reflection in cooperative learning based on similar diagrams by Harvey et al [117] and Lucas [164]. Critical reflection of the cooperative education experience depends on the combinations of a student's experience in university and the workplace.

a Venn diagram, with the intersection of the areas of the university learning, work experience and the student being the point of reflection. This seems a valid model. The student's experience (including prior to the work experience), goals and ambitions, along with their knowledge and understanding, when combined with the university academic experience (from classes, instructors and advisors), as well as the work environment that includes supervisors, projects, colleagues and general work environment all can combine in a reflection. The depth of reflection can vary from student to student, some may not progress to deep and metacognitive levels and instead stay at a low level of reflection by recording what was done. But the process of reflection on specific work experiences could be returned to over the years and perhaps (unwritten) reflection will be carried out be the learner as they consider their work placement and the impact it had on their learning. It should be noted that Lucas' diagram and model considers a more transactional process occurring between the learner, university and work provider, with different aspects moving into the reflection and out from the various elements. The nature of the transaction between the three entities is worth consideration.

The return to the campus after a period in industry may provide a different perspective on the courses taken and their approaches to the courses. A period of working and seeing what is needed in the work place may reveal areas of shortcomings and aspects to focus on. The familiar environment and process of taking classes, laboratories and studying in the university may be looked at with a different perspective. How different can vary from student to student and is dependent on each individual's cooperative learning experience.

As we can see from the above description of the cooperative workplace learning experience there can be a significant number of aspects at play in a situative model to the working environment. The different stages, from preparation, work and reporting the experience, there are different influences on the learner. Those can be the environments of the workplace and the university preparation, to the people who are guiding and working with the student and the type of work the student is doing and what they are learning technically. Overall the complexity can result in a rich and rewarding experience, that can influence the future career of the student and sometimes significantly influence it.

7.4 Chapter Summary

In this chapter we have looked at one form of experiential learning where faculty in the home department of the students do not have much direct influence. The students' learning is in the hands of the co-op employer. Historically is appears that engineering employers are keen to help and be involved in the education of students, who may be future employees. The benefits to both employer and student are there and it can result in employers finding perspective employees through the coop process, or at least helping to prepare students for professional employment in engineering or otherwise. There is evidence too that there can be benefit for students in gaining employment and earnings following graduation.

The learning that occurs in co-op can lend itself well to being examined using a situative learning perspective. Working on employer projects, with experienced engineers and applying knowledge from university as well as learning new knowledge, to bring back to the campus, can be viewed through the lens of a situative framework. The reflection on such a key experience can also enhance the learning.

From a personal perspective I can appreciate the benefit of co-op education. As a student I had an opportunity to work for a manufacturer of optical fibre. This experience and work was valuable enough that this led to later summer employment, as well as full-time work upon graduation. Later I returned to university to do a PhD, still with the area of optical fibre, but moving from testing fibre to undertaking computer simulation work on understanding fibre lasers. The benefit of having handled fibre and measured some of its parameters, allowed me to at least relate to the material I was simulating. Later when I was working in the communications industry I had no hesitation as a manager to have co-op students within my team. I even hired some to work in my group after their graduation. As a former co-op student and employer of co-op students I saw the benefits first hand. I have also understood the importance of helping the education of co-op students.

Before moving on we should consider the situation of students who do not do co-op. They will likely have a continuous period of study and any relevant work experience will be limited to work before starting studies, summer work, or part-time work. This work, although useful may not be as effective as a co-op position, where the focus is on the student gaining long-term experience in employment. So for any students who do not have relevant work experience by the time of graduation they will be missing on some of the benefits that co-op students will have gained from their work, that has been outline in this chapter. It will take a year or two of employment to bring up some of the knowledge and experience a co-op student has already obtained. There is the key benefit that co-op experience can help with securing a position and the possibility of it affecting starting salary [276].

8

Beyond the Curriculum. Undergraduate Research and Student Societies

8.1 Introduction

So far we have considered experiential learning that is part of the curriculum. Even cooperative education usually gets recognized on the transcript. However, there can be important and relevant learning that can occur on a campus that is often not part of a course or a program. This learning can be when a student is working in a faculty member's research group, or working in program relevant student societies or engineering competitions. This type of experience may be recorded in a university's co-curricular record or similar, if they have one, but at the least, it can be recorded in a resumé or CV. Working on a problem in a research project team can help a student learn more about their discipline and subject area, as well as gain insight and experience into what it is like to conduct research. Participating in an engineering society or an engineering competition can help develop organizational and leadership skills, as well as engineering and design knowledge. All these types of involvement can develop the student's self-identity in an engineering context. In this chapter we will look at this valuable form of experience that can occur beyond the curriculum, though we do note that some universities may have formalized working on a research topic within a course, such as with a capstone project (already discussed in Chapter 6).

If we consider the role of a university to break down into three areas i) to create knowledge and understanding, ii) to disseminate knowledge and the related understanding and iii) to assess knowledge and understanding; then we know that most undergraduate education is deeply immersed in ii) and iii). What then about including students into i), which is typically the area that graduate students and faculty are focused on? The Boyer report in 1998 proposed that the US research focused universities to make changes and adopt ten proposals, Table 8.1 [240]. The first change proposed was to 'make research-based learning the standard'. Within this was the proposal to involve undergraduates in the research process, to mentor every student and to include internships.

DOI: 10.1201/9781003007159-8

TABLE 8.1
Boyer report's suggestion for undergraduate involvement in research [240].

1	Make research-based learning the standard
2	Construct an inquiry-based freshman year
3	Build on the freshman foundation
4	Remove barriers to interdisciplinary education
5	Link communication skills and coursework
6	Use IT creatively
7	Culminate in a capstone experience
8	Educate grad students as apprentice teachers
9	Changing faculty reward system
10	Cultivate a sense of community.

By the time of the Boyer report there was already an established organization to encourage undergraduate research in the US, the Council of Undergraduate Research (CUR). It was founded in 1978 and has grown considerably since then and had an established and well-attended conference, the National Conference of Undergraduate Research. A conference where undergraduates from across the US and beyond can present their research work.

With organizations like CUR and reports like the Boyer report, there has been a steady development of funding and support within universities for undergraduate research. One common model for undergraduate research is to provide support for undergraduate students to work directly within a faculty member's research group.

Another area that engineering undergraduates can get significantly involved with is related societies and clubs, such as student chapters of professional societies, as well as in student engineering competitions. These competitions could be organized through a student group or through an academic unit. Examples of the competitions could be the racing of student built Formula SAE cars, the Simulink® student competition, organized by MathWorks®, and the Great Northern Concrete Toboggan Race, which has been running annually for over 40 years in Canada. Whether it is an amateur radio club, or a competition where a vehicle is being designed to race against other similar designs, there is engineering learning in the organization of the varied technical aspects and running of the club or activity.

One key factor is that generally these extracurricular activities (although related a student's program) are voluntary and involvement is because of the interest of the student. Being outside of the main curriculum, but linked to the discipline of the program these experiences can provide valuable opportunities for learning.

8.2 Undergraduate Research

Before we take a deeper look at undergraduate research (UR) it is perhaps worth considering what it is and what makes it different from other forms of student learning. One of the leading organizations for supporting and encouraging this form of experiential learning is the Council on Undergraduate Research, which was formed in 1978 and is based in the USA [269]. They have an Undergraduate Research Definition Task Force who provided a definition to be:

> "A mentored investigation or creative inquiry conducted by undergraduates that seeks to make a scholarly or artistic contribution to knowledge." [268]

This provides a inclusive and succinct definition. The 'mentored investigation' indicates that the undergraduate is supported and guided by another with suitable experience in the field and inquiry methods. The 'contribution to knowledge' indicates that there is likely something new or additional to be added to a field and that it implies that there is the communication of the information found as a result of the research. The use of 'creative' and 'artistic' also allows the inclusion of fields involving design and the arts.

The form that undergraduate research can take can then be broad, whether involving activities and investigations in a research laboratory or a classroom, and be implemented through one-to-one guidance, problem-based learning or project work. For examples and discussion of successful implemented approaches of undergraduate research, the book 'Developing & Sustaining a Research-Supportive Curriculum: A compendium of successful practices' is recommended [146]. As we have dealt with problem-based learning, project work and innovative classroom-based approaches elsewhere, in this section we will mostly restrict the examination and discussion of undergraduate research to the form of one or more undergraduates being involved with a researcher and possibly a research group. The researcher is the guide to help solve a research-like questions and help them learn about research methods and to help them through a way of reporting the findings. The process of research is often something that is learned in graduate school by many and to include undergraduates requires giving these students help and guidance with the full process.

If undergraduate students are to participate in research they usually need to receive something in return, beyond the experience, whether that is academic credit or financial compensation. This is dependent on how the undergraduate research is incorporated into a program. Formal incorporation into a degree program most likely gives the student credit towards the program. Alternatively, the research may be part-time work or part of an internship or some other program. The organization, funding and support of these opportunities can come through different routes. Some may be organized at a national level, through funding bodies that provide supporting monies for training undergraduates in research. The motivation here is to not only help the student to get to try research (and if they like it they may go onto graduate school), but also to help support the researcher who is perhaps funded by them. One scheme in Canada is the Undergraduate Student Research Award (USRA) that helps faculty already funded by the National Science and Engineering Research Council of Canada to hire an undergraduate student during the summer term when most programs are not running courses [51]. Other levels of government may also support, such as state or provincial level. Along with this government level of support there may be local institutional programs put in place, at university, faculty and department levels. Ash Merkel reports on the development of undergraduate research at four research universities in the US, including MIT which started its Undergraduate Research Opportunities Program in 1969 [15]. In the paper it is noted that faculty are usually eager to involve undergraduate students in research and this becomes an educational opportunity for undergraduates. They are able to work with leading researchers and a chance to work within a community of scholars that can include postgraduate students and postdoctoral fellows [15]. The culture of undergraduate research is noted in the paper and the enthusiasm for UR and its development can extend from administration, through faculty to students. UR scholars can be involved in seminars, undergraduate research journals, research clubs and poster sessions.

How do faculty consider undergraduate research and the inclusion of undergraduates in their research teams? A paper looking at faculty perspectives of 155 faculty at the university of Delaware on undergraduate research noted that in 85% of cases the research collaboration was initiated by students [291]. This may be an indication of the process for supporting undergraduate research at that university (there was a supporting unit), but it indicates a strong interest by students. 90% of full-time faculty include undergraduates in their research, which shows a strong support for UR. The program at Delaware was reported as having a summer research program and a senior thesis program too [291]. Time spent supervising the students was reported by 50% of respondents as 1 to 2 hours per week, 40% provided 3 to 5 hours and 10% spent over 5 hours per week. In the survey 75% of faculty indicated that influencing the "career of talented young students" was a important or very important. Whereas 50% indicated that contribution to their research was important or very important. 41% thought that it was important or very important to have this contribute to "the quality of life" at their university [291]. This shows a strong support for influencing the student's careers, and perhaps trying to attract them as graduate students. It is interesting to see having the students contribute to the faculty's research was a lower motivation.

As for how the undergraduates were integrated into the research the survey showed 91% of faculty had graduate students involved with the undergraduate. This may provided added supervision as well as helping graduate students develop their teaching and leadership abilities. The adjustments made to include the undergraduate work was listed as:

> "1) create smaller problems;
> 2) assign exploratory problems;
> 3) integrate undergraduates into existing projects
> 4) devote designated time and money to the undergraduates." [291]

These adjustments can help accommodate the undergraduate students because of the short duration the students may have with the research group (significantly less than a graduate student) and the likely reduced technical experience an undergraduate student has compared to others in the research team.

What about the impact the research experience has on the student? Zydney and co-workers also assessed the impact on alumni of the University of Delaware [292]. In doing this there was a comparison with students who had not had a research experience as an undergraduate. The survey had 245 respondents with 37.1% being identifiable as having been a participant of the institution's Undergraduate Research Program; 26.9% self-identified as having undertaken a form of research as an undergraduate; and 35.9% of respondents had no undergraduate research experience. When comparing the responses related to skills and abilities from the questionnaire there were some significant differences between the responses from those that had done the Undergraduate Research Program and those that had no undergraduate research experience. These differences had a higher rating from those who had done the research program and the skills and abilities were [292]:

1. Speak effectively.
2. Understand scientific findings.
3. Know literature of merit in field.
4. Analyze literature critically.
5. Possess clear career goals.

The training and experience from the undergraduate research may help the students to become familiar with engineering and scientific literature, and hence the higher response values for items 2 and 3 in the list. The fifth item, clarity in career goals, could be linked to the experience of the research as the authors also noted that over 80% of those had respondents who had done the Undergraduate Research Program had finished, or were underway with, a graduate degree program, compared to less than half of the those that had not undertaken an undergraduate research experience. As well, over 50% of those that had been in the research program or had reported having had a research experience, reported that it had been a faculty member had been 'important' or 'very important' in influencing them to attend graduate school. The paper's authors clearly note that the response did not show if this faculty influencer may have been the research advisor or not, it may indicate that the closer interaction may be a factor.

The communication ability, listed first above then becomes an interesting outcome, which is perhaps related to the research experience. In another study examined a larger population of engineering students, 5126 respondents across 31 engineering schools, and specifically looked at the impact of undergraduate research experience on communication, teamwork and leadership [52]. These areas being graduate attributes examined in accreditation. This analysis of the data from this large survey revealed that there was no significant impact on the teamwork and leadership skills, but did show there was "a significant predictor of communication skills". The development of communication skills may then be related to presentation of work and findings as part of the research work. Presenting within the research team, as well as possible presentations of the work at institutional events, or even national events such as the National Conference of Undergraduate Research (NCUR), can all be an important way that student presentation skills can be enhanced. Is it surprising that teamwork or leadership skills have not been significantly developed beyond what is expected? If one considers that in a research group students will have perhaps the least experience and will be looking to others to advise and guide them, then it is not unexpected those skills are not significantly developed. The skills and abilities development situation will be different in the next section when student groups and engineering competitions will be considered.

Studies of science students have also looked at the impact of undergraduate research. For example, Seymour et al report that after interviewing 76 students from four US liberal arts colleges there were reported benefits in an number of areas, including (reporting in descending order) [239];

> "personal/professional gains (28%), "thinking and working like a scientist" (28%); gains in various skills (19%); clarification/confirmation of career plans (including graduate school) (12%); enhanced career/graduate school preparation (9%); shifts in attitudes to learning and working as a researcher (4%); and other benefits (1%)." [239].

The 'various skills' mentioned included communication, especially presentations and arguments, laboratory or field techniques, reading comprehension and collaboration [239]. Although this was for science students it not unreasonable to think that there may be some commonality with engineering students. With this science study and the earlier mentioned engineering studies it can be seen that the reported impact of undergraduate research can lie with the help on career paths, including graduate study, as well as communication skills and no effective evidence of leadership development. The study of science students and their recognized benefit of 'working like a scientist' is notable from a situated learning perspective. Specifically here students are noting that they a feeling like they are becoming scientists. Again it is not unreasonable to suppose that engineering students may feel similar. As the science students feel a sense of working like a professional, then this can enhance the identity of the student and that they are directly moving towards becoming a professional. Like co-op work, seeing they can do the work required in the profession and that they can work effectively alongside others, does show that enhancement of situated learning and moving towards the goal of being able to work in the profession.

If we look at the situative learning of undergraduate research further we can consider the three areas of action and interaction, mediation and identity. The type of specific approach to undergraduate research may vary, dependent on whether the research is funded for a summer (as full-time employment), as part-time employment throughout a year or as a thesis or independent study course. Opportunities can vary dependent on what programs and routes are available, for example Davis provides details of what programs have been available at University of Colorado, Boulder [72]. These include an undergraduate research opportunities program, independent study (as a technical elective course), senior thesis and a Discover Learning Apprenticeship. This latter is a university wide initiative to match undergraduates with faculty research projects.

Working in the research team may be everyday, for a period of time, or a few times a week. There may be space in a laboratory, or in an office, for the student to work. There will be regular contact with the research supervisor and often there can be contact with other faculty, post-doctoral fellows and graduate students. Each may provide help and support with the research work. Plus, there may be presentations to the research team too. The student undertakes an apprentice type role with the research supervisor guiding and the others in the research team becoming colleagues and advisors as needed. Hunter et al, took social constructivist approach with their examination of faculty and student views of undergraduate research in science [133]. This is follow-up work on the previously mentioned investigation by Seymour, et al [239], that includes more perspectives from faculty that are involved in undergraduate research. The authors note that from the data the faculty can be better at interpreting the development of the student's abilities as new researchers and scientists, than the students themselves. The students may recognize their own learning, but it may take an experienced external observer to see the overall development in an undergraduate student learning to be a scientist. This work is based on studies of science undergraduates it seems reasonable to assume the findings are applicable to engineering students too.

In undertaking the work there will be use of equipment in the research group. This could be lab equipment, or simulation tools. There will also need to be reading of relevant literature and documents. Mediation with these tools and literature will be needed to develop the knowledge for the research. As the project is underway there could be a need to design, evaluate, plan and make purchases. All will require the processing of information. The training and the analysis of the findings will be done with the supervisor and other researchers. This sharing of meaning of data or literature with other members of the research team brings the student further into the community of researchers and away from the starting periphery of that community [133].

As for identity, we can see from the earlier mentioned research that the research experience can impact the decision to go onto graduate school. Many students may undertake undergraduate research specifically because of an interest in graduate school and to see if they are interested in research. Hunter and colleagues mentions the chance to 'try on' science research taken by science undergraduates undertaking UR [133]. Seymour et al found the impact of doing research caused students to recognize they were now working as a scientist [239]. Again, this is likely to also occur for engineering students to undertaking undergraduate research. Working closely with faculty and research team on a specific research project will be a significant step in an undergraduate's perception of their progress to becoming a professional. They are very likely being paid for this work too and will be coming onto the campus to not just undertake classes, but they are in a lab or research group in a work environment. That work could include co-developing in the project or demonstrating what they are doing. They will also have to present work to colleagues in the research team and possibly to others beyond the group. This employment may be their first, or one of their early positions, that is technical and related to engineering. Students may have done a co-op position too. This experience may be influential and strongly adjusting the student's identity as an engineer in training.

8.3 Case Study

Activity: Undergraduate students join a research group
University: Tokyo Institute of Technology, Tokyo, Japan

At the Tokyo Institute of Technology a large percentage (approximately 90%) of undergraduate students go on to graduate school [254]. Students start to gain experience of working in a research group in the 4th year (although students in the Transdisciplinary Science and Engineering Department join in the fourth quarter of the third year) when they join a professor's research group. This involvement can provide undergraduates with study space access in a research laboratory, attendance at the research group's internal seminars (usually weekly) and also the registration into an independent research project course (for credit). The choice of which research group the students join is based on their interest, program of study and current grade point average. Having joined a research group the students now have a base for their studies. They also have exposure to other students in the research group who are undertaking post-graduate work. Allowing undergraduates to engage with masters and PhD students provides experience and insight as to what is involved in research in their selected field. They also get chance to see the research work of others develop as well as formulating their own research work. The research groups may have visiting students and scholars from overseas too.

Interactions with co-students occurs in shared space in research rooms, as well as in weekly research seminars. Each week there is a review of what has been achieved since the last update by the students in the group. Newly joined 3rd year students do not usually have to present their progress initially, but when they reach their 4th year they should have a research question decided with their supervisor and they will regularly present updates. Direct supervision is provided by the supervising professor, but graduate students can provide some advice too. So a broader support group for the student is provided. Not needing to present after first joining the group allows the student to see what is entailed from the other researchers in the group.

Students undertake their research and their final assessment is a research paper, submitted to their supervising professor at the end of the research. The student passes the course on successful assessment of their report. If a student stays at Tokyo Tech and progresses onto a master's degree, there can be some movement between research groups. In some departments it is more common that in others.

From a situative learning perspective the students are working in a dynamic environment alongside others at different levels of research. There is access to research-level resources and the shared office and laboratory space allows a community to develop. The students can observe and see different active research projects. There can be a development of identity as time is spent in the research group, becoming familiar with the environment, all while the student's own independent research skills and project are developed.

The key features of this approach to the independent research project is the exposure to research and research methodology. The students see over a year the development of other student's research as well as their own. Sometimes assistant professors are involved too, working with a more senior faculty member's research group. Seeing the example of the others can assist the undergraduate student's own work and can also help them decide if they do want to undertake post-graduate research degrees. The model seems to work well for Tokyo Tech as their transfer rate from first degree to graduate school is so high. The size of the student population makes the inclusion of the students into laboratories and to provide them with desk space. There are about 10,000 students (the includes both undergraduate and graduate) at the university at the time of writing this and the campus has three different sites around Tokyo.

8.4 Engineering Societies and Competitions

Besides spending time in program studies there is also the advantage for students in college and university to join student clubs and societies. These can be sports, political and specific interests related. They can also be linked to their studies. For engineering schools there are usually a number of clubs and societies related to engineering and technology. They could be directly associated to a discipline of engineering and specific programs, they can be a general student society for all forms of engineering or they could be involved in a specific form of technology, such as amateur radio or artificial intelligence and these could be open to students of any discipline. These societies offer a form of learning and deepening of knowledge that goes beyond the curriculum, but which can take advantage of the facilities and support within the university. Organization and projects undertaken are done by the students, sometimes with a faculty advisor. As well as the general and local operation of the clubs and societies, there can be local, national or international competitions that students can be involved with. Even though all of these society activities are outside of formal program studies, they are done on the campus, often with the help and support of faculty and each can provide experiential learning. Although, sports and other societies can bring there own learning and possible paths to careers, this section will be restricted to considering engineering, technology and science-related societies.

Students join clubs and societies for a range of reasons, but mostly usually it is due to interest and a sense of belonging. The society may give them a chance to develop an established or new interest and skill. It can be a way to meet others with similar interests and backgrounds. There may be access to informal space and to technical equipment. It can be a way to join a smaller community which can provide help and support. It can be also a way of joining a larger professional community as many professional engineering societies have student chapters along with a student membership option. For example, the IEEE, SAE International, the American Society of Civil Engineers (ASCE) and the American Society of Mechanical Engineers (ASME) all offer a student for a low fee. This allows access to the society that they may stay with throughout their professional life, as well as allowing the student members access to resources and possible local chapters.

The student societies may be interdisciplinary and if recognized by the home engineering school and/or the university, it can provide access to funds and other support, such as common rooms for work and connecting with fellow students. Alternatively, there may be more of a connection at a departmental level and a society is strongly connected to a specific discipline or degree program. Again, space and funds may be allocated. Then there may be groups focused on particular activities, such as the previously mentioned amateur radio club or artificial intelligence club. The advantage here is the access to expertise of other students, possible graduate students and faculty. An amateur radio club can help support training to obtain an amateur radio licence, as well as access to a radio station with transceivers and antennas. Other societies may be connected to students and campuses to support engineering students and engineers from minority groups in engineering such as the National Society of Black Engineers, Women's Engineering Society or the American Indian Science and Engineering Society. These societies can provide support, networking and mentoring within an institution, locally and nationally.

What of the general activities of a club or society? These groups are generally student lead so there will be leadership positions within the structures. These positions will help guide and organize the group and its activities. So students can gain experience in leading, organizing projects and possibly handling budgets and space utilization. If the society is technically focused there can be hands-on projects. I had the benefit for being the advisor of a student amateur radio club. Example projects undertaken were construction of transceivers from kits, the building of antennas and the measurement of their parameters, the installation of temporary antennas, as well as more general operation of the radios to make contacts locally and beyond, including in other countries. Students can get a feeling of achievement using an operable radio station they constructed and set up themselves. The work can provide more practical application than perhaps encountered in their program's course and the making of radio contacts is a form of reward that is highly enjoyable and memorable. There would be similar technical learning and use with societies that are for example focused on car construction and racing, unmanned airborne vehicles or concrete canoe designs. In Wankat's paper on undergraduate student competitions there is a quotation from a surveyed faculty member who says:

> "Being able to conceive, design, fabricate, construct an object, and then see it perform in action is an invaluable experience for young engineers. It helps them develop a sense that engineering is not just practiced 'on paper'; it is making their ideas work in a physical sense." [278]

Although this quote is directed towards student engineering competitions (more on that later in this section), the comments about the practical aspects and the leadership skills are also valid for some technical non-competition-based societies. As there are no grades to consider and worry about there is no academic credit impact. The reward is to see the item work and then perhaps improve on it. There is an opportunity to technically explore.

The experience of organizing and obtaining the funds for the activities provides a depth and ownership of the activities that goes beyond what may be experienced in a course or even a co-op position, where there would be little control on the type of work or task assigned. As mentioned in the earlier quote, this type of management and leadership experience will be attractive to potential employers.

Besides the projects there may be organized meetings, guest speakers and visits to companies or other institutions. These will provide members a chance to be involved with a community of people with similar interests but also with a chance to get more engaged with the interest and find support and knowledge.

As mentioned engineering competitions can be a part, and a major one, of some engineering student groups. They usually involve the construction of some item, such as a racing car [137], a concrete canoe [14] or some code for a particular application [174]. Then there is a competitive assessment against other teams and submissions. Organizers can be professional societies, companies, or both may partner. Participants may not need to be in a specific club and instead they may be linked to courses, including projects. Regardless, the activities usually require extra work, outside of conventional classes, for the preparation and participation in the competition. In the case of large projects, such as vehicles, there can be considerable logistical organization to travel to competition venues, again developing and testing leadership skills.

So how popular is participation in engineering competitions like this? Simmons et al have considered the out-of-class activities of engineering students and found that engineering design competitions were the third top activity [241]. Table 8.2 shows the list of the top six activities. It can be seen from the table that the design competition team comes close to the second.

TABLE 8.2
Out-of-class activities of engineering students, based on responses of 649 participants of a survey from three US institutions [241].

Out-of-class Activities	Percentage students
Job	16.8%
Sports	12.5%
Design Competition Team	11.9%
Culture, Faith, Gender, and Identity	8.5%
Professional Experiences	6.8%
Research	6.0%

In 2005 Wankat reported on an analysis of a selection of engineering competitions to see if there was a consistent winner for such competitions [278]. Although his findings showed there were regular winners in individual competitions it did not appear to be the case across a variety of competitions. When looking at the reasons why, it did appear that dedicated faculty advising was part of the reason. It was also noted from survey responses that the tradition of winning, the resources available and the quality of participating students were all additional factors. The question of the impact of competitions on learning, and whether students' learning improves from participating Wankat notes that this was harder to answer from his survey of faculty advisors. There are benefits (see the quotation above from the survey), but it was noted that some competitions could be bad learning experiences. Schuster et al also reports on the benefit and challenges [233]. With regards to the challenges they note the impact of the competition timetable on the design, forcing compromises in the design. There is also the impact on classwork for those students who become focused on the competition design and spend a disproportionate amount of their time on it, or if they have to travel to a competition during term time. The constraints of resources are noted too, which if too significant can be demoralizing to the students. Other factors can be seen to be changes in design for the sake of changes, or building without designing. There is also the disappointment when there is failure in the competition and the risk of embarrassment. The role of the faculty advisor becomes important to help guide students (perhaps to attend classes!), but especially with respects to safety. Both Wankat and Shuster et al note the importance of the role of the faculty advisor and though rewarding in many ways, such a role can have a considerable impact on the advisors own time and they too have to maintain a balance [233]. The level of involvement, especially with decisions has to be measured, so that the advisor does not intervene or take over too much. There is the advice "to leave the design decision-making to the students" [233]. Another issue can be related to the gender imbalance. With the high percentage of male students in engineering, this can have an effect on the gender ratios in the engineering competitions and societies. Simmons et al shows from their survey data that design competition teams are predominately male and favours the racial majority [241]. The impact of gender in engineering competitions has been examined by Foor et al in a qualitative study [99]. This study has some revealing narratives and shows the challenge of getting into the competition team and shortcomings of working within the team, including little assistance in learning. Part of a transcribed response from a female student who had turned up on a Saturday for a CAD workshop, a subject she had never done before. She goes on to say in the transcript that organizers of the workshop gave her something to build and then left. She believed it was a "trial by fire" [99]. To help avoid such issues Foor and colleagues recommended three things for advising faculty [99]:

- "Understand the peer group cultures at your institution."
- "Interact in informal face to face ways to recognize and correct any negative manifestations of peer culture."
- "Provide appropriate and effective mentoring to develop team leaders and teachers."

Despite the potential for problems there are significant areas of benefit. Being involved in a design and build that has its performance challenged in a competition can be valuable for the students. There is the learning of real-world issues such a learning about planning, procurement and lead times of parts, budgeting and documentation. Some builds can also be done over a number of years. If the project is outside of a course and in a student club or society then there is the potential for starting work in first year on a competition project and staying with it over the years until graduation. This allows the seeing of the evolution of a design over a longer period than even capstone projects. This also can help the student develop their role into leadership within the competition project.

Leadership has already been mentioned as a learning benefit from participation in the team competition, but also as a necessity to provide appropriate mentoring. The impact on students' leadership identity development through working in student groups that participate in engineering competitions is examined by Wolfinbarger et al [285]. This study used a leadership model by Komives et al [152] using six levels to categorize leadership, see Table 8.3.

TABLE 8.3
A model of leadership identity development. Simplified from a model by Komives et al [152]. Used by Wolfinbarger et al to student's leadership identity. [285]

Stage	Brief outline
1. Awareness	Child has recognition of leaders and authority figures. Others are leaders.
2. Exploration/engagement	Experience of interacting with others with similar interests. Perhaps recognition of their own leadership potential. Generally like a high school student's perspective.
3. Leader identified	See leadership as a position. Looking to leaders for direction. Considers people as leaders and followers. Many entering tertiary education could have this perspective.
4. Leader differentiated	Seeing leaders as what individuals did and not only due to a position. If in a position of leadership see their role more as a facilitator.
5. Generativity	Creative to care for others and sustainability of the group. Responsibility to group members, groups and does mentoring.
6. Integration/synthesis	Self-aware as leaders. Do not need to have a position to engage in leadership. Recognize they can and need to learn from others and open to self-development (life-long learning and commitment). Confident to work in groups effectively.

As competition groups are self-motivated in the main the development of leadership within a group can be important to help it continue and succeed, though there are faculty advisors who can assist with this. The study Wolfenbarger and colleagues undertook was a qualitative approach to answering a set of research questions on leadership identity, done through interviews with officers of an engineering student teams that competed in engineering competitions [285]. From this they determined from the volunteers that there were leaders at leadership stages of 3 through 6 with some showing evidence of transitioning to the next stage. The authors give brief quotations showing the students operating at the different levels. One at level 6 is aware that leadership is a journey of self-development and they recognized they were a good leader but wanted to develop further. That student also understood how to work with others, including groups, to affect change. Whereas others showed not such a developed sense of leadership, such as one the author's evaluated at level 3 who had a response of:

> "If you can't get people together and tell them what the goal is and when we're going to do things, it simply won't happen.... The goals of the team are always going to be set by the captain." He doubted the members' internal motivation: "Most ... show up to do something fun or to put something on their résumé. It's the captain and maybe a few of the leads who really have to actually push if we want to really reach for something." [285]

The study gives insight into how students view their leadership and other student's leadership. Awareness of the six-level leadership development model may be useful for any faculty members who act as advisors for student clubs and societies, and more broadly in activities like projects where leadership in students can be called upon and developed.

The awareness of the benefits to learning by participating in professional societies and engineering competitions is known by students too. This is discussed in a paper by students from the IEEE branch of UNED, the Spanish University for Distance Education [170]. Here the perceived benefits of their student society and the link with a professional society, in their case the IEEE. As well, this is an interesting case as the society student members are distance learning and spread across Spain, so this student branch helps link the students outside of the virtual classroom. Of professional societies the authors note that the societies become a way to transition from university to industry. Participation takes a student beyond formal education. And of student groups they say:

> "But in these groups usually learning does not happen spontaneously. Some kind of impulse is required to encourage students to participate in an active way. Usually, it happens through the participation on contests (e.g. robotics, programming, networks, etc) or through interesting projects to achieve a particular goal." [170]

They go on to note that beyond contests, participation in technical talks or publication of papers can also be a motivator for students.

This indicates the student's awareness of the value of what the professional society can offer, as well as the need for projects and competitions. The paper details the activities of the student group including robotic workshops, involvement in the Cubesat project and national and international congresses. Activities which include support from the professional society.

This leads us to consider the role of student engineering societies and their associated activities from a situated learning perspective.

8.5 Situative Learning Perspective of Student Engineering Groups

As can be seen in the involvement student's have with student societies, some of which can include professional societies and engineering competitions, can produce a different learning environment from course-based learning, or the earlier discussed undergraduate research. If we now consider this from a situated learning viewpoint, we can consider the three aspects of *action and interaction*, *mediation* and *identity and participation*.

8.5.1 Action and interaction

To join a university club or society is a voluntary decision for the individual student. Once joined the student could be a long-term member and remain in the group until graduation. Participation in the group can involve attending meetings and assisting with the group's organization and events. Membership can develop into specific roles and involve leadership and mentoring newcomers. Engineering societies may be involved with workshops and technical learning. Guest speakers may be invited to deepen knowledge in the group's area of interest. Projects and engineering competitions may become a central focus of the group. Advice may be provided from one or more academic advisor, but in general a group's direction will be guided by the students. Like many aspects on a campus, there will be a cycle of events based around the academic year. For a group this can start at the beginning of an academic year with a new leadership and recruitment of new members, moving through the projects and events of the year until towards the end of the academic year when there will likely be elections, or otherwise, determining of who will be the leadership next academic year. Senior members will graduate and members who were new at the beginning of the year will have accumulated a year's experience. Some students may leave, even though they have not yet graduated. Projects can become multi-year projects and improvements and development can continue year after year. There may be space provided by the institution for the group and their projects. This provides a central base and place of belonging. The space may have technical equipment. Funding for the equipment, projects and any related travel will have to be found by the group. This may come from the university or from professional societies. Developing a budget, bidding for funds and accounting of the monies will be part of the group's work and will likely follow an annual or termly cycle.

8.5.2 Mediation

Engineering projects within a group may be practical and involve construction. Students will work with tools and knowledge of the use of tools may be passed on through the group, or garnered from academic courses students take. Some tools may never have been used before. Plans, diagrams and software may be used within the group. These may require study and training. Simulation tools may be used and developed. Terminology and technical language will be learned. If there is a competition then the rules and requirements will need to be understood and followed. Some projects may require external licensing and so understanding and meeting the licensing requirements will need to be done.

8.5.3 Identity and Participation

Students join a group and may find they know little about the details of the group's work and activities, as well as members of the group. There will a connection with a different group of people to their program studies cohort. Some students will be of a different year and will have different levels of experience in the groups interest area. Students will have roles and responsibilities, in a considerably different way to courses which are led by instructors. There may be a faculty advisor, but they have a different role to when being in a course. As part of the membership to the group students may join a professional society. This provides the students with access to professional resources and potentially opens up networks of other members. There may be meetings and connecting with similar groups from other institutions and if involved with competitive projects there could be competition and rivalry between the groups. A competitive aspect can lead to a stronger sense of group identity and sense of purpose. There may also be the pressure to succeed. An individual student may rise to named roles and acquire responsibility within the group. Leadership, whether in a named role or not, may be expected and developed. The sense of purpose and responsibility can go further than that encountered in many academic courses. Students may also interact with external groups and individuals. Networking with external engineers and other professionals may occur and these may also be potential employers. This connecting with professionals, as well as design and engineering projects can develop the engineering identity of individual students within the group. The process of successfully undertaking a role within the group can further strengthen the student's engineering identity. Failure at competitions or with projects can also adversely impact the student's engineering identity. The interaction with the faculty advisor can also be different to that encountered as a course instructor. The advisor may become more of a mentor and guide.

8.6 Chapter Summary

In this section we have looked at areas that are often outside of the curriculum and have faculty in an advising role. Students participating in undergraduate research or engineering societies and competitions will have chosen to be involved in the activity, unless the research activity is a mandatory part of the program. So this motivation will be out of interest, the challenge, or perhaps to develop further knowledge. If the undergraduate research is a paid program then there may be the additional motivation to earn money.

Involvement with the faculty member will be as an advisor. With undergraduate research the advisor will be as a subject expert. In an engineering competition or a student society the role of the advisor will be less supervisory and more advisory, and there will be more interaction with other students. Learning from students can occur in both societies and in undergraduate research. In societies it can be from other undergraduates and often from those who have been in the society for a number of years. Whereas in undergraduate research the students are most likely to be graduate students. Both areas can provide experiences that cannot easily be encountered in a typical course. The active participation in a society and any related engineering competition can last a number of years, and that could involve a number of competitions. The ability to participate through a competition team in an design, a competition and subsequent redesign is a real opportunity to see the evolution of an engineering project. Whereas in undergraduate research there is the ability to work on current problems in research where the outcomes are unknown. The direct participation in the research process can help develop a deeper knowledge of a particular area. There are similarities to undertaking co-op employment however the research is usually still in the university environment and involves research questions and literature.

From the discussed literature it can be seen that students undertaking these two areas can develop different knowledge through the experience. In the societies and competitions the area of leadership and design experience can be some of the key areas of development. Whereas in undergraduate research there can be technical knowledge development, experience of using research literature, as well as communication skills. The experience with research can also clarify whether a student wants to go onto graduate school or not.

What is clear with both of the areas considered in this chapter is that participation in one of these areas can be a strong and worthy experience for a student. The benefit of the extra knowledge, and often practical knowledge with equipment and processes, be helpful for the student in ways that a classroom-based course may not provide. A critical factor is for a student to know what opportunities exist and the support they receive from advisors. Student societies are usually well advertised and there may be a strong word of mouth. Undergraduate research can depend on provision and support, financial and advising, within the university. A limited level of support may restrict the availability to a few and often the high achieving students, if the grade point average is used as a key selecting factor. This can be an incentive for students to achieve high grades, but it does not help with including a significant number of students in research activities. Having a variety of different programs in a university, faculty or department may provide options for becoming involved in research.

Both of these areas are ways that students can less formally interact with advisors and learn directly from them. It also allows the advisors to know the students better, perhaps provide guidance and to help provide informed references for potential employers or graduate schools.

9

Lessons from Other Professional Programs

9.1 Introduction

In recent years there has been a growth in interest in experiential learning across different programs in higher education. In some subject areas experiential learning has been established for a while, such as co-op programs, practicums and laboratory work. In this chapter we will look at some of the experiential learning in professional programs outside of engineering to see if there are ideas and experiences that could be applied to engineering programs, but currently are uncommon or not found at all. In an attempt to limit the wide choice we will focus on three disciplines that provide professional programs: medicine, business schools as well as social work. These three areas have similarities to engineering, such as accreditation to expected standards, routes to professional recognition and oversight of professions by external organizations. Of course there is experiential learning in disciplines and subjects outside of the professional programs but for this chapter we will limit our choice to these. Also, the examples given in this chapter from the three areas are samples. The examination is not meant to be a thorough survey, which would go beyond the space of the chapter. Instead these are selections intended to show how experiential learning has been implemented outside of engineering, but may provide ideas that could be implemented in engineering or other disciplines.

Within this chapter we will also look at one area of assessment that has been strongly adopted in medical education, which uses an observed performance in the assessment. This type of assessment is called objective structured clinical examination. Rather than having the assessment based on written responses the student has to conduct an actual technique, procedure or observation in front of an examiner under standardized conditions. This becomes a clear assessment of experiential learning and practice of the procedures or techniques. This assessment approach is well established and used in medical professions and similar but has had limited use in engineering, although it could have potential application.

DOI: 10.1201/9781003007159-9

9.2 Medical Professions

Traditionally the medical profession has used modes of experiential learning, with work in hospitals or clinics, diagnosis training, laboratory work and practice with equipment, treatment and procedures. However, like other areas of education in recent years there has been an increased use and application of experiential learning in medical education [289] along with the use of reflection [229].

In this section we will give some examples of reported experiential learning that may inspire ideas for engineering education. These will look at developing knowledge, learning by doing and placing the students in an immersive situation for learning more about the people they will work with.

Many practitioners trained in Western medicine have little knowledge of the complementary and alternative medicines. In one project, a training program was set up by an academic medical centre to introduce cardiologists to different types of alternative medicine [124]. The Chair of Cardiology approved the release time of the physicians to undertake the program. The training comprised of five hours of experiential workshops to learn more about six different alternative medicines including yoga, Tai Chi and meditation and three hours of lectures to provide details and evidence of the alternative practice. This was to help the physicians support and council patients who may be interested in or using complementary and alternative medicines. A study to look at the training was undertaken, and included a control group, that focused on looking at knowledge, attitudes, likelihood of changing practice patterns and influences. The results showed that the cardiologists who attended the program had more understanding and favourable attitudes to the complementary and alternative practices. There was also a request for more research information and details about the alternative medicines used. A benefit for both the doctor's knowledge and their patients who may be interested in using the alternative medicine. The program was subsequently offered to other specialist medical groups [124].

Here the report illustrates the use of experiential learning to educate practitioners in topics that they would not normally have considered, though their patients may have requested information. As the types of treatment were not from Western medicine the practitioners may have not been as receptive to the approaches without having experienced them. Experiential learning can then be an effective way to inform practitioners and guide them towards more understanding in areas that they may have not have familiarity and even perhaps preconceived views. It is worth speculating whether a total of eight hours of lectures would have helped inform and educate the course attendees as well as the five hours of experience and three hours of lectures, that were used.

Medicine requires a range of technical procedures and most will require a degree of skill and dexterity to carry the procedures out. In some ways this is similar to engineering although there can be more pressure on time and a risk with failure in medicine. One report has shown how experiential learning was used to train medical students to undertake endotracheal intubations and was compared to the use of a method that had been traditionally used at the medical school [259]. The procedure is used to keep the airway open in a patient, but it is not without risk [259]. Practice of the procedure is needed to maintain competency. In this work the training of the procedure was done using a manikin and the study looked at how successful students were when learning using either a traditionally used step-by-step guide to performing the procedure, or a truly experiential learning process of learning to do the procedure by trial and error with the manikin, without any training. The interesting result was that the group of students who learned by themselves were 64.5% successful in undertaking an incubation 3 months after the training, compared to 36.9% of those who learned from the step-by-step guide. At 6 and 9 months after the initial training the results of both groups were similar, but the authors report at 12 months the experiential group had increased success with the procedure compared to the step-by-step learners [259].

The third and final example here is a powerful example of experiential learning, the Learning by Living Project [112] where students were immersed into a real situation for a significant period of time. As part of their medical training students were given a diagnosis, confined to a wheelchair and became a temporary resident in a long-term care home for two weeks continuously for 24 hours a day. As a resident of the nursing home the students learned about what it is like to be in a home under the care of others. They lived with the regular residents and were given similar attention by the care staff as provided to the long-term residents. The students also had to keep a reflection journal and snippets are included in the paper [112]. The value to the students is they not only observe but they experienced the day-to-day lives of those in the care homes. This extended to details, such as how visitors respond to those in wheelchairs, or the need for carers to come down to eye level of those they give care to. The learning was not just by the students, but the care home employees and management also received feedback from the student. That helped shape and change some policies and procedures, along with the way individuals deal with those in their care.

This last example deviates from many other experiential learning approaches where the student takes on the role of practitioner. Here the students are immersed (for two weeks) in the role of a care home resident and the students are the recipients of the professional practice. Their lived experience will shape their approaches to those needing long-term care to a depth that a lecture, seminar or brief visit is unlikely to provide. How this role reversal could be applied in an engineering setting is worth considering. The example has the students in the long-term care home has the students in the form of a customer to the practitioner. This could be the obvious route for the engineering student to experience and learn. For any application or development work, if possible, the student could experience what it is like to be the user or customer. One key feature that comes out from the Learning by Living Project is the duration the students are in the role of resident. This ensures the students are exposed to a full range of experiences, not just a 24 hour period but two full weeks. The reflection becomes the way that thoughts and experiences are processed and considered. So this is not the expert guiding the novice but the novice learning from first-hand experience. The result is perhaps a significantly deeper appreciation of the challenges encountered by those needing care.

9.3 Objective Structured Clinical Examinations

Assessing students is perhaps still most commonly done through written examinations. These are useful for the assessment of knowledge-based course content, but if the course has experiential learning then assessment can take the form of reports, reflections, presentations and similar as mentioned in other chapters. In the medical professions there has been developments and use of a series of standardized practical examination to test diagnosis, psychomotor and interpersonal skills amongst other competencies relating to practitioner interactions with patients. This has become known as objective structured clinical examinations (OSCE) [148, 177], and has been adopted in a number of countries and in a number of medical disciplines and beyond, including for police training and evaluation.

The OSCE originated in the mid-1970s with the aim to better assess undergraduate medical students' clinical abilities [113]. The aim of the OSCE is to test an individual in a number of scenarios to see how the individual performs. The performance can be assessed and graded to give summative or formative feedback. The student starts the test and proceeds to an assigned test station where a situation is presented to the student and they have to undertake an examination or test. Then questions may be asked or the procedure is observed and after a few minutes, typically 5 minutes, the student being assessed has to move onto another station and a different situation or scenario. Examiners observing at each station can make a structured assessment based upon a grading sheet, or the student answers questions at a nearby station, unable to return to the assessment area. The student continues around all the assessment stations, without encountering a station for a second time or returning to a stations for a second attempt. Prior to this approach, clinical examinations used patients on a ward with one or more examiners following a student and observing their clinical skills. As Harden et al pointed out the structuring of the examinations at stations, along with the prepared marking scheme ensures that variability of the patient and examiner is reduced (a problem with the original clinical examination), so the only variable in the assessment is the student being assessed [113]. Because of the manner of the assessment. The weak results of a performance in the OSCE can indicate an all around poor performance, or it could indicate there are specific deficiencies if results are low in only one or a few areas. Hence it can be used for both summative and formative assessment. This control and feedback has ensured the adoption of the OSCE approach.

OSCE can be considered for the assessment of programs as well as student assessment. Here students can be tested with an objective structured exam to see if the outcomes of a program or course are being met. This could be useful for accreditation purposes.

Use of OSCE in the medical profession covers a wide range of disciplines, see Table 4.2 (p. 46) in [115], which includes dentistry [169], nursing [226], radiology [189] and veterinary medicine [121]. It has also been used in England and Wales for promotion purposes in the police force [182]. So far there appears to have been little use of objective structured examinations in engineering, though details of an approach to assessing practical skills was reported in 1980 [114]. One report on a pilot use of an objective structured technical examination (OSTE) has been reported [5]. This had interest and support from both the students who were examined and the faculty who delivered the examinations at the station. This attempt to apply an objective structured examination was done with first year and second year electrical and electronic engineering undergraduate students at the University of Hertfordshire, UK, who were assigned a series of 16 tests devised by faculty. These tests were set up at stations and included a range of short tests such as, soldering components, setting up an oscilloscope, measurement of a waveform period, frequency and peak-to-peak voltage and finding a fault in an op-amp circuit. When surveyed [5] 88.9% of students rated the experience of the tests helped their confidence, whereas only 52.2% of students rated the sessions as very good or good on a five point Likert scale. When the faculty were surveyed 90.6% rated the sessions very good or good. Students wanted to experience the sessions three times a year whereas the faculty only two times. When asked as whether to use the OSTE in the curriculum 93.1% of the assessors compared to 53.5% of students were in favour. When asked if the OSTE should be used as part of summative assessment the assessors were 75.9% in favour whereas only 28.3% of students were.

It is interesting to see the support that the faculty/assessors give the OSTE. With the students there appears to be more support for the use of the OSTE for learning and perhaps formative assessment. The immediate evaluation, under a time pressure, in the presence of the assessor may make students less likely to want to have the OSTE for summative assessment, than perhaps a written examination. However, it does appear that students recognize the benefit of having their more fundamental and practical skills assessed. The assessors meanwhile seem to see a clear benefit of this form of assessment, and having this in a repertoire of assessment techniques is deemed valuable. The merit of such objective and structured testing has been recognized widely in the medical disciplines and has been used in assessing students. It appears from this pilot that there is potential in having similar testing in engineering, although there is no subsequent detailed reporting of this type of testing being implemented, that I can find. With the increased interest in experiential learning as well as the rise in outcomes-based learning there appears to be merit to exploring more of this form of assessment in engineering, which has become so established in the medical professions.

9.4 Business Schools

Business students undergo similar types of education to engineering students. Subject and disciplines are often compartmentalized into courses. There are numerous quantitative and analytical subjects. There are ethical requirements and future employment is often in a business setting that requires the knowledge of the latest developments and regulations.

Like engineering, business students are being exposed to more experiential opportunities, not only with the more traditional co-op and internship employment, but also in the classrooms in business schools. In this section we will explore a few examples to see how experiential learning is being implemented into business schools.

It is common to find case studies in business school courses. Although these a routed in real-world situations [168] and students can study and reflect on them, as experiential learning they can lack a direct experience and so miss that key stage in the Kolb learning cycle. One initiative to provide that direct experience in a business school was to use job shadowing [179]. Here in a relatively brief way (over around 8 hours) a student could follow a business leader or manager to gain insight into the role, profession and challenges faced. This could be an element of a course, in this case it was for a business communications course [179]. The experience of the shadowing was followed up with required reflection exercises that involved a journal and a presentation. Within the job shadowing the students conducted a survey of employees on their jobs (such as requirements and satisfaction). The processing of the data from the surveys turned into a research activity that involved the writing of a report and multimedia presentation. When the students in the course were surveyed the job shadowing element of the course was cited as the most helpful element of their work, with the next two categories being the speeches and the tests. Case studies were ranked fourth in that list.

The role of the business librarian in helping with information literacy in three different experiential initiatives at the University of Nevada Las Vegas (UNLV) are discussed in a paper published the Journal of Business and Financial Librarianship [109]. The three type of experiential learning are field-based consulting initiatives, student competitions and student-managed investment funds. These are different types of activities but each involves collecting information and applying knowledge to an authentic and real-world challenge and in most cases dealing with businesses and groups outside of the classroom.

The field-based consulting described by the author details three different courses at UNLV, from first-year undergraduate to MBA level, where companies could seek assistance from the students who would act and consultants. These ranged from first years providing ideas to senior management of Domino's Pizza on how to increase lunchtime revenue in their outlets, to business students working as interns to helping local manufacturing companies develop export plans to global markets. With these student consultancies there is a reciprocal benefit to the companies and the students who are experiencing direct learning as well as linking this to their course learning. One consultancy that linked to engineering students was having the business students consulting on engineering capstone projects. Here the business students could help engineering students by looking at the capstone projects and advising how to turn them into businesses.

Other examples of experiential learning provided by Griffis is the investment fund managed by students and student competitions [109]. The investment fund management is a learning situation that requires up-to-date information to make informed decisions for effective management of the investment portfolio. The student competition is a co-curricular activity that allows students to competitively apply and test their course-learned knowledge to a business challenge. This is similar to the engineering competitions that have been described earlier.

The involvement of the business librarian in the work described in the paper reveals another strand to the support for the experience that goes beyond the obvious course instructor and the business that provides the challenge. The partnership with the librarian provides a level of professional support that aids the learning and helps the students to undertake the challenge. In engineering there may be similar support from the library as well as from technicians in workshops and laboratories.

A further business case we can look at involved a classroom-based experience looking at business ethics in marketing. In the example students are led through a three pass loop of examining an ethical dilemma scenario. The use of the loops of learning is to reinforce the learning through evaluation and reflection of business ethics, including the presented dilemma, in different ways. The students were also provided with a model on ethical decision-making and they had to read this prior to being presented with the case scenario [132]. The authors of the report note that although some students acknowledge understanding of the model after reading it though often this is found to be at a superficial level. The inclusion of the case study is done to show the use and reinforce the model to the students. So the first learning loop is initiated with the understanding of the model. At the beginning of a class the students are asked to read and consider an ethical dilemma scenario, which in this case was a sales manager discovering some of his salespeople were providing cash gifts to purchasing agents. After being presented with the case the students are asked to form their own opinion of three potential solutions that are given to them. They must rank them as most to least ethical and can use the model they have been given, or follow their own moral views. The ranking of the students' responses were collected. Next the students discuss why there were different views across the class. This stage completes the first learning loop. The start of the second loop consists of examining the scenario again with perspectives from the model. This stage includes an examination of the range of stakeholders and how they could be affected. Solutions from using the model were then discussed and assessed. This closes learning loop 2. The final loop is opened when the students are asked to reflect on their own moral codes and how this affected their previous analyses of the scenario. This reflection completes the whole exercise [132].

9.5 Social Work

Social work is another profession like medicine that has clients that need their services and has rigorous training to provide effective help and solutions to the varied issues and problems. Clients may need careful and measured interactions with practitioners to determine what assistance is needed to the individual, who may be in a state of crisis. Like medicine, social work educators have made use of experiential learning to help the students become effective practitioners. As in previous sections we will look at some published examples of experiential learning used in social work education.

One study that has extended the Kolb learning cycle from four stages in a cycle to a five point approach, shows how authentic issues can be used in a role-play situation [64]. The extension of the Kolb model was to include a fifth stage of 'formative integration of the whole', which was to act as stage where the other Kolb stages were considered holistically, to help with understanding the process from examining the issue, through to helping provide a solution. The authors state

> "In clinical practice, learning is cumulative and formed through diverse practice modalities that examine techniques from the various roles: client, professional, case-owner, supervisor, and active learner, for the integration of knowledge and practical skills. The use of this formative integration concept will add insights to help learners gain practical education from a variety of perspectives." [64]

Here the authors acknowledge the need for the practical-based learning to be considered from a variety of perspectives. To facilitate this multiple perspectives, the four stages of Kolb's learning cycle are assigned to a role within the social work environment. Namely, *active experimentation* links to the social worker; *concrete experience* - the client; *abstract conceptualization* - the clinical supervisor; *reflective observation* - the case owner. The added fifth stage of *formative integration of the whole* links to the social work learner. One other variation added to the Kolb cycle is the recognition that using this five point model in social work the flow between the stages is not cyclic but is can move between the various stages (and perspectives), although the added *formative integration* stage cannot be considered until a few of the other four stages have been visited.

The study that contained this five point Kolb approach involved a role-play type session in a clinical practice course that immersed students in scenarios that had come from the students' personal experience [64]. This use of the students' personal experiences created an authenticity as well as an objectivity when the individual observed their own case. As can be expected there was care applied with the confidentiality of the information supplied by the students, including review by the institutional review board at the university and the instructor being a licensed clinical social worker. The students provided details of a personal issue, taking on a client's perspective and met the instructor having done a three-generation reflection on their own family. A case was then written by the student, taking on the intake worker role, and the instructor then changed details such a race, gender, age and ethnicity for confidentiality. Collected cases from the students were then passed out students who were paired. Two cases each were passed to the students one for client and one as clinician. By delaying the release of the pairings and encouraging no rehearsal the students then could move to the simulation and role-playing the client and the clinician in a case simulation. Both students in the pairing undergo different experience, one experiencing the role of the social worker guiding the case investigation and the other as a client reflecting on how it feels to be a client. Other students observed behind a one-way mirror or via a closed-circuit TV as either designated supervisors or observers. The student observers were to provide immediate feedback, with the 'supervisors' required to link the case and solutions to the theoretical aspects taught in other courses. Knowing all the cases came from students in the course provided a known authenticity to the overall simulation, albeit with confidentiality of the source of the case. Each student had a chance to observe their own case being role-played. A closure session was provided, along with the opportunity to see the instructor individually to discuss anything arising from the sessions.

This example, like the earlier medical case where the students spent two weeks as a resident in a long-term care home, shows the role-reversal aspect of being a client. Whereas the long-term care home resident experience was a deep immersion as a client, this example shows the student to be required to have multiple roles, as a client (both with their own personal case and the assigned case), a social worker, supervisor or as an observer. This mix of roles with authentic cases provided a rich experience that could ensure the students process and consider integrating the various experiences in different roles holistically, so completing the fifth stage in the author's experiential model of formative integration. Reported feedback from the students was favourable.

Another study of social work students taking the role of practitioners is reported [242]. This time practitioners were participants in the interview roles and the students taking on the role of the interviewing practitioner were early stage social workers. The role-play session was recorded so students could see their body language, mannerisms as well as interview style. There was also a reflection required by the students.

The evolution of a particular approach to include experiential learning is documented by two Faculty members in a social work department at the University of Manitoba [41]. In this work students use photography to document aspects of lives to provide a 'photo voice' approach [253], a form of participatory action research. This report details the three-year evolution of a social work research course. In the first year of the photovoice project the students of the course were the focus and after learning about qualitative research theory they undertook the assignment to document their "Life as a Northern Student" using photographic images. These images were then shared and reflections were written on them.

In the next year of the course there was an expansion of the range of the project by linking to a community beyond the campus. The students were briefed on the theory, ethics and privacy relating to photovoice and photograph taking. The project then was named "Photographs Generate Knowledge: Northern Manitoba Homeless Research Project" and community agencies and a local heath centre helped bring the students and participants together. The photographs taken were analyzed with the participants. The third year was run in a similar way to the second year, but this time the students worked with an agency that provided temporary accommodation for families and individuals. The project looked at near homeless.

In each of the three years the students learned ways to apply qualitative research through the use of imagery, with the photography being used as a way for the social work students to see the experiences and life of others. The authenticity of using the community partners added a further dimension which students noted and is recorded in the paper.

The two examples in social work described so far have looked at role-play and research-based approaches to experiential learning. Both have offered an immersive experience as well as trying to at look situations from another′s perspective. Since social work requires the ability to see and understand the challenges and predicaments in other's lives this exposure can only help create the awareness in the student practitioners. As each student has their own perspective it is valuable to have multiple ways to expose the students to the different social and social justice situations. One other way is to use simulations or games. There is no doubting the success of games and simulations. The game gives a scenario or challenge and the rules provide a framework by which the challenge or test can be overcome. In this third example from social work a simulation was given to students to understand the challenges and opportunities to a community development due to wealth, or the lack of it. A game was run for graduate students during orientation week at the Boston University School of Social Work [96]. The game split the students into the roles of different socio-economic groups and then encouraged them to build their ideal community with the resources and supplies that they were provided with during the game. The socio-economic groups include an affluent group (that had half of the building space), a middle-class group (with 30% of the building space) and two lower socio-economic groups (that had the remaining 20%) of the space. Faculty and administrators played other roles in the game including chief of police, police officers, housing authority officials. Unknown to the students these other roles distributed resources unequally and treated the groups differently, favouring the affluent area and being unfavourable to the low socio-economic groups. Over the 45 minute duration of the game there was an unequal development of the communities.

As students will come from different socio-economic backgrounds this simulation provided a focused and clear example of how inequity can not only manifest itself but be systematically perpetuated. The effect, as with the other social work, examples was for individual students to see beyond their own life experiences, in this case to look at social injustice. The authors conducted a survey of the students and also provided a discussion session after the simulation. It was noted by the authors that there were regular discussions of discrimination and the discussion provided a venue for students to challenge one another's opinions on this. Finally, it was noted by the authors that the simulation did help with illustrating issues surrounding socio-economic class privilege.

9.6 Chapter Summary

In the previous sections in the chapter we have looked at some experiential learning approaches from other disciplines and professions that are perhaps less common or not used in the engineering discipline. As mentioned in the introduction these selections were not intended to be comprehensive, nor representative, but were intended to illustrate different ways that educators in other professions have been creative to encourage learning in their students. Table 9.1 includes a brief summary of most of the cases described for the three disciplines covered. What can be seen in each of the cases is a clear experience for the student that has likely a much stronger impact than would likely have been obtained with a traditional lecture. Some of the experiences may be brief, such as the socio-economic privilege social work game or significantly longer like a two-week long total immersion in a long-term care home for medical students. Each does involve a period of reflection for the student. Many of the studies report that students rate the experiences and learning highly.

TABLE 9.1
Summary of activities from other professions described in Chapter 9.

Discipline	Experiential learning activity	Key aspect
Medicine	Alternative medicine education	Education in a less traditional area for doctors
Medicine	Intubation training with a manikin	Practice to maintain competency
Medicine	Learning by living project	Assuming role of patient for two weeks
Medicine	Objective structure clinical exams	Performance assessment in a simulated live scenario
Business	Job shadowing	Observing and reflecting on the role of a business leader
Business	Being consultants	Students become consultants to a business
Business	Ethical dilemma	Applying an ethics model to a scenario
Social work	Role-playing	Role-playing client and clinician in an authentic scenario
Social work	Photographs generate knowledge	Use of photographs and imagery to see others experiences
Social work	Socio-economic group game	Game simulating effect of inequality

Perhaps obvious, but worth noting, is the time invested in creating the learning exercises by the instructors. In many cases this is time to organize and setup the experience, as well as debriefing. Because of the nature of the exercise the student's time spent in the experience has to be considered, as too long spent in the exercise could affect the time to be spent on other topics and exercises in the course.

Re-examining Table 9.1 we can look at the type of exercises listed. In some cases the activities experience what it is like to be the professional, such as conducting an endotracheal intubation, job-shadowing or being a clinical social worker in role-played simulation. Alternatively, some of the exercises required the students to adopt the role of clients to the profession, such as: being a patient for two weeks in a long-term care home; participating in alternative therapies or role-playing a client in a social work simulation which had come from a real experience of another student in the class. This 'tables turned' approach makes for a very interesting alternative approach to experiential learning. Considering how such learning activities and experiences could be created in an engineering setting is perhaps worthwhile doing. Spending time as a 'user' of a prospective product or item to be designed and built could be very valuable. This does not have to be the end user of the final product, but it could be as next recipient in the design or construction stage. I recall a story from a designer colleague who explained how he changed the location of a bolt fixing, in a device he designed, after he saw the assemblers struggling in the process of assembling the units and even hurting their arms. Perhaps having students extensively test project designs, acting as a user or even quality control could be valuable to see design or engineering deficiencies. Similarly, students could develop product requirements and specifications for other students. This may happen in multidisciplinary group projects, when students working on one aspect of the project have requirements from a team working on another aspect. This 'internal' user is common in industry.

The use of job shadowing (from the Business examples) could be applied to engineering. This could be considerably shorter than a co-op experience but still give valuable insight into roles and potential careers. Few engineering students may have a clear idea of what a typical day is for a project engineer, field engineer or a quality engineer. These experiences, coupled with reflection after the experience could help with relating other studies to the work of the shadowed person (facilitating a form of abstract conceptualization perhaps).

The use of games and simulations too within engineering courses could be innovative. This could be a challenge but could be valuable in course such as professional practice, including ethics [157, 247], business and management [27], or engineering economics. A survey study of games in engineering education has been published in 2016 [39], looking at the published use of games (both electronic games and otherwise) in educating engineers between 2000 and 2015. This work highlights the overall perceived benefit by authors of the surveyed literature, but sees the need for more systematic study of the efficacy of the use of games in learning.

The final observation to made concerns the objective structured clinical examinations. This is one of the few situations where a performance and experience becomes part of an assessment. This has become well-established in the medical field. Given the laboratory and practical nature of engineering this type of examination could be used in assessing the abilities of engineering students. Despite the limited application to engineering, and consequently the limited studies of its use, here does seem to be one area of experiential learning and assessment that could be used in engineering schools, especially with outcomes-based assessment. It makes an interesting example of the potential lack of cross-pollination of ideas between disciplines. Here then is a final point of this chapter, that much can be learned from other disciplines, whether they are doing similar approaches to ones in engineering or whether there is a fundamental difference. Educators, regardless of discipline, are all looking at authentic and appropriate ways to educate their students and future practitioners of their profession. Researching, observing and sharing these ideas across the disciplines can help all our students.

10

Engineering and Society

10.1 Introduction

If experiential learning is to connect education and students with the real world then being aware of current areas of issues and concerns of society becomes important. Engineering as a profession is involved with global, environmental and societal issues, and not just technical. In this chapter we will look at a few areas where there has been a demand and need for change. Topics that may need to be considered within the engineering curriculum and within the engineering practice and the profession. For example, in this chapter we will consider diversity in the engineering profession, as well as decolonization which affects countries that have a colonial past and finish by briefly considering some ethical issues. Beyond the education environment if engineers are to be meeting their professional obligations to society they need to have an awareness and appreciation of current concerns.

It should be recognized that different countries will have different issues. Such countries with a colonial past will need to deal with the effects of colonialism and the impact on Indigenous peoples. Many people in the countries that undertook colonizing may not realize that this issue is being grappled with by colonized countries such as my own country of Canada. Similarly, countries around the world may have different discrimination issues-based around race, gender, sexuality, neurodiversity and ethnicity. Concerns that centre around equality. Going into this chapter there are some issues that may not seem as relevant to the reader as others. However, much can be learned from how engineers are helping to tackling these issues and remove barriers. It is important to make the next generation of engineers aware of the challenges and attempts at solutions, so as to have engineers aware of societal issues and to be involved in solutions.

In the 1955 Grinter report (reprinted in the appendix of [116]) on the then 'Evaluation of Engineering Education' made a recognition of the obligation of the professional engineer to society and recommended ensuring students are aware of this. After identifying that engineering education's first objective is to ensure students are to be technically proficiency in their discipline,

DOI: 10.1201/9781003007159-10

the report identifies a further objective involving leadership, ethics and general education. This second set of objectives can can be seen to be including what is needed for the professional work and practicing in society. Indeed, the report further mentions to include "an understanding of the evolution of society and of the impact of technology on it" [116]. The report also encouraged the adding of humanities and social studies courses to the curriculum of engineering programs, supporting the ASEE's then call for it to be a fifth of the curriculum [116]. Today most engineering programs contain courses on ethics and having humanities and social studies courses in the curriculum of engineering programs.

There is some irony with this support for 'the understanding of the evolution of society' as a modern reader may find the report's occasional reference to students as 'men' surprising. For example, when referring to the quality of teachers it says (on page 76 of the reprint [116]) "To achieve these goals he should possess energy, enthusiasm, and a sincere interest in the development of men." Showing an assumption both the instructor and students are men. This may be indicative of the times when the report was written and the then assumptions that only (or mostly) men would become engineers.

Today engineering faculty are no longer 'male-only', neither is the student population, although in most countries it is still overly represented by males and this has been an identified issue. The UK Government's (supported by the Royal Academy of Engineering) Perkin's Report highlighted the UK had the lowest number of female engineering professionals in Europe, at just under 10% from 2007 data [207], p. 15. The highest reported percentage in the report was 30% in Latvia.

This diversity issue for some countries can go beyond gender and for some nations it can be the mismatch between the diversity of the racial origins of the general population and the diversity of the population of engineers and engineering students. Going further, in some countries with colonial pasts there can be a strong mismatch between engineers of settler, or immigrant, origin and engineers who are Indigenous. This extends into a lack of recognition of the value of Indigenous knowledge in engineering schools. How then do engineering schools teach other ways of knowing and how are Indigenous knowledge keepers involved?

As technology and innovation develops there can be resulting challenges. The acknowledgement of climate change due to human activity requires engineering, amongst other professions, to help reduce impact as well as to try and solve the problems. Experiential learning work could include aspects of sustainability, to help prepare future engineers. Similarly, the development of technological solutions that require decision-making in a potentially dangerous environment is highlighting ethical issues that engineers are and will face. Autonomous vehicles are one area that has been recognized [225] where decisions from sensor systems have to ensure safety for passengers as well as other road users and pedestrians. Recent high profile examples such as the Volkswagen emission case [12] and the Boeing 737 Max crashes [106] have highlighted how engineering decisions can draw in ethical problems [175]. If engineering students are to be prepared and exposed to learning in the real world awareness and education of the ethical issues should be made.

In this chapter there will be an examination of some of the societal challenges that affect engineering and engineering education. These could impact on experiential learning, as well as provide opportunities for learning about professionalism and professional practice. The sections in the book will deal with diversity, Indigenous contributions and ethics. Each of these sections will provide preliminary details that could help an educator who is undertaking experiential learning. Each of these areas is worth deeper study beyond what can be given here.

10.2 Diversity

It would be reasonable to expect that the diversity in engineering should be similar to that of the general population of the country. However, in many countries that is not the case and for many years there have been attempts to redress this imbalance and have better representation. If engineering is to help all, it again seems reasonable to expect the profession to be as diverse as the population it serves. This ideal certainly is challenged by a legacy of social and cultural practice. In many countries engineering may be stereotypically seen as a profession for males and although there have been changes away from that opinion it may persist. Whether consciously or unconsciously, female students that expressed an interest in science may have been guided away from engineering through stereotypes, towards other areas like the health sciences or architecture. In many Western countries the majority of students entering will be male and most likely within that group the many that will be what could be labelled as 'white' or Caucasian. The lack of racial diversity may be due to stereotypes, socio-economic reasons and the lack of opportunities in the past and currently. The impact though is a lack of inclusivity and the risk of perpetuating the stereotypes of who can be an engineer.

The American Society of Engineering Education collects data from a number of US engineering schools and publishes it in a report, "Engineering and Engineering Technology by the Numbers" [85]. For the 2019 report the percentage of women graduating from engineering program was 22.5%, with the men accounting for 77.5%. The survey has now started to collect data for non-binary or other gender, but they report for 2019 that "The numbers collected this year are too small to draw many conclusions from". So less than a quarter of engineering graduates are female, whilst the US Government census data reports the percentage of women to be 50.8% for 2019. The percentage of women graduating has increased continuously over the prior 10 years, with the percentage being 17.80% in 2010 and 19.10% in 2014. This is considering the overall number of engineering graduates and these percentages may be different within specific degree programs. The '... by the Numbers' report does look at the gender distribution across different disciplines or programs can vary. Table 10.1 shows a sample of the data of the percentage of women by degree discipline collected by the ASEE for the 2019 report (including the highest and the lowest values) [85].

TABLE 10.1
Percentage of females graduating from engineering disciplines with degrees. Reported in [85] from surveyed US institutions.

Program	Percentage women students
Envr Eng.	51.7%
Biomedical Eng.	48.1%
Chemical	35.8%
Engineering (General)	29.9%
Civil	25.4%
Engr. Science and Eng. Physics	21.4%
Mechanical	15.7%
Electrical	14.4%
Aerospace	14%
Computer	13.3%

If these distributions have been relatively consistent over the years, it can be assumed that a US co-op student will be more likely to encounter a female engineer in some disciplines over others. As well, most disciplines have predominately male engineers working within them. This lack of potential role models and mentors can slow moves to more of an equitable balance in gender diversity.

ASEE '... by the Numbers' report also details data from a few Canadian schools. For bachelors degrees in engineering, from eight Canadian engineering schools, it was reported that 25.2% are female. So a little higher than the US, but still about half of where it would be found in the general population. Other countries will have their own breakdown and it may be valuable for an educator to look at their own country's population diversity distributions and compare to those of engineering graduates in their home country or school. As mentioned in this chapter's introduction, the Perkins report (published in 2013) details the percentage of female engineering professionals in Europe. The UK is reported as having the lowest percentage amongst 28 European countries at just under 10% in 2013 [207]. Across different countries the number rises including about 16% for Germany, about 20% for the Czech Republic, 26% for Sweden and the highest being 30% for Latvia. The 2015 UNESCO 'Science Report: towards 2030' gives a table showing percentages of female tertiary graduates, including for engineering [270]. This shows more globally the variation of women graduating in engineering, the lowest being Saudi Arabia, 3.4%, and the highest being Myanmar at 64.6%. Table 10.2 shows a sample from the UNESCO report.

TABLE 10.2
Share of female tertiary graduates in engineering using data from 2013 or closest year. Sampled from [270].

Country	**Percentage female graduates**
Argentina	31.0%
Bahrain	27.6%
Ghana	18.4%
France	25.6%
Malaysia	38.7%
Mozambique	34.4%
New Zealand	27.4%
Tunisia	41.1%
Turkey	24.8%
USA	18.5%
Viet Nam	31.0%

Apart from two exceptions in the UNESCO data (Myanmar and Oman) all are below 50%. These percentages show the variation is country or regionally dependent. The reasons could be complex and could in part be due to population, industry, social and economic factors, as well as past and present cultural norms.

The overall low participation rate of women engineering students means they will likely be in the minority in courses and experiential learning can involve groupings of students and that means that women can be in the minority. One study looked at the experience of Japanese women engineering students at two universities in a laboratory setting [126]. With the women graduates being about 10% of the engineering graduates in Japan, the study showed that being in a significant minority could lead to experiences with their male student peers that could be discouraging and affect the students confidence [126]. They would often be the only women in groups leading, to a feeling of token status. The study found feelings of exclusion by the students, including in study groups outside the lab. Participation was affected, either in deference to men, or not wanting to be judged which impacted on the assertiveness of some of the women. For some, the challenges lead to a decision to leave engineering. Although this study was specifically for Japan, it is not unreasonable to expect that similar challenges are faced by women engineering students in other countries.

The challenges of women who have been practicing engineering is reported in a study that looked at the experiences of late career or retired women engineers [87]. This work can give some insight into potential challenges that women engineering students may encounter in industry on coop placements, or after graduation. The engineers in the study had graduated in or around the 1970s in North America, which was a period where the number of women was very low at the start and had risen slowly to the present levels. What is noticeable is the issues encountered by the practicing engineers were similar to the students in the Japanese lab study described earlier. The three common issues in the responses from 251 surveyed were the challenges:

1. not getting respect from their peers and supervisors;
2. not fitting into the work community, including being isolated and left out;
3. obtaining a work/life balance, especially with family commitments.

Again, the issue of exclusion and being judged by male engineers arises. The working engineers reported "subtle biases to overt discrimination" [87], and having to not only understand organizational dynamics at work, but male organizational dynamics. Similar to the students, the challenge encountered in the working environment caused some of the engineers to leave their work altogether. When asked about how the situation had changed over the time since they started working, the overall response from the engineers was that "progress was slow".

For an instructor in a class of engineering students what then should be the lessons from these studies? First, given a gender imbalance there can be challenges with how students in groups interact. Besides the normal difficulty of getting to know a stranger who becomes a fellow group member, there is an issue for female engineering students who may be in a minority or even a single female in the group. Steps should be made to help overcome any group dynamics problems. Hosaka recommends instructors to provide "guidance, monitoring and feedback" [126]. The feedback may be needed by some to help overcome self-evaluation and risking a lack of confidence. The UNESCO 'Cracking the Code' report gives similar recommendations on helping to build self-confidence and self-efficacy where needed [270]. Building on this, both Hosaka [126] and 'Cracking the Code' [270] recommends creating inclusive environments so all students can work collaboratively. If there is a lack of inclusion in lab or project courses then this can mean that the wrong behaviours are learned in school and then carried over to the work environment.

The data of the breakdown of ethnicity [85] (using the report's naming of the groups) for non-international graduates in 2019 is shown in Table 10.3. This table also includes the US population estimates for 2021 from the US Census Bureau [48], for comparison to the general US population.

For the US Census data more than one race can be declared and so the total for that column exceeds 100%. As well, for this table the "White alone, not Hispanic or Latino, percent" was included.

TABLE 10.3
Percentage distribution of the ethnicity for non-international engineering students in the US, along with the general US population for comparison.

Ethnic background	**Percentage engineering graduates**	**Percentage population**
White	60.8%	59.3%
Asian American	14.7%	6.1%
Hispanic	12.1%	18.9%
Black or African American	4.3%	13.6%
Unknown	3.8%	
Multiracial	3.7%	2.9%
American Indian/Alaska Native	0.4%	1.3%
Native Hawaiian/Other Pacific Islander	0.2%	0.3%

When considering the literature equity, diversity and inclusivity (EDI) in engineering work the connection to engineering identity arises. Who is considered to be an engineer and what constitutes being an engineer can bring in stereotypes of what historically and conventionally an engineer looks and behaves like. For the general public in many countries this will be a male, and for most Western countries a white male. Pushing stereotyping further is could be someone who has a technical and scientific fascination, and who could colloquially be called a 'nerd'.

Moving beyond the stereotypes the need to understand engineering identity becomes important when we recall the situated framework (see chapter 2), for considering how a learner develops their knowledge through a community and its practice. As well, if there is an expected way an engineer is to appear and behave, how does a prospective engineer react when they see they do not look or behave like the expected and traditional engineer? It becomes clear that the expected norms can then be a deterrent to diversifying the engineering profession and making it more inclusive to all. As the decision whether or not to study engineering can come in the teenage years for most then the perception of the engineering identity can not only attract people to the profession but also deter those that might consider it. Some suited for the profession may not see themselves fitting into it. This then can perpetuate the stereotype as the perceived image acts as a filter.

As we have seen with the data of the participation of women, mentioned earlier in this section, change can then be slow. For examinations of literature on engineering identity Tonso [262, 263], Morelock [190] and Rodriguez et al [222] can be starting points. As Morelock points out in his survey 25% of the studies on engineering identity he reviewed were connected to studies of engineering students from underrepresented racial and ethnic minorities. This was closely followed by studies on women and gender. Moving beyond students, one study that was focused on identity and practicing women engineers is reported by Faulkner [92]. This is a UK-based study in engineering related to buildings and building operations and was conducted through job-shadowing. It looked at gender in engineering, gender in engineering identities and boundaries between engineers and architects, as well as between engineers. The work of the civil engineers involved not only technical-based aspects but also social-based work, when it came to managing people and projects. Hence part of the title of the paper, 'Nuts and Bolts and People' [92]. The study examines the role split of the work, between a technical or 'technicist' work and a social or management work. The resulting gender connotations of the two aspects; technicist being viewed as male and the social/management being female is explored. The spectrum of the work of an individual along the technical-social scale then has implications on how engineers may view other's work, their own and their career progression. The term 'gender authentic' is raised and can relate to how technical work and technical prowess could be viewed as 'real engineering'. The social and management type of work could be viewed as requiring and using more general prowess and with a view that it is 'only' management [92]. The division is sometimes not so clear as when management roles can link to 'money power' and the idea of 'marketplace manhood'. The complexity of the genderization of engineering work can clearly affect perceptions of engineering identity, as well as influence career directions and progressions.

So why is gender or ethnic and racial diversity a concern in experiential learning? Within a school setting experiential learning very often involves working with others. If there is a lack of diversity then it is possible individuals could feel less accepted or understood and this could lead to exclusion or prejudice by others, which should not be tolerated.

Experiential learning exercises can often be challenging, so there should be no unfairly added burden. To have to deal with being excluded can be an unwanted challenging prejudicial experience. For experiential learning in a work place, such as a co-op term, the lack of diversity can also lead to inclusion challenges, but as well there can be a difficulty to obtain mentoring or identifying with a role model, who understands some of the diversity challenges the student may be encountering.

Educators will not be able to provide the solution to diversity in engineering on its own. However, there can be steps taken to help with making experiential learning a safer space for all participants for all. These are suggestions and advice for consideration and may be dependent on the activities involved in your experiential learning.

- Look at your own positionality. Your own past, current position (including power), identity and views can affect how you consider others. Perhaps you come from a privileged group and have not faced discrimination that others encounter. Recognizing your own background, identity and perspective may be valuable and may help to consider challenges and positions that others may have, who have a different background to you. My own positionally will be given in the following section on Indigenizing engineering education.

- If you employ group work consider carefully how groups can form [224]. If students are allowed to form their own groups there can be a risk that the diversity in some groups could be low. We have seen in the above data for the US that the most common gender and ethnicity of a graduating engineering student are male and white. If a random grouping is made then there is likely to be a broader spread and as well friends and pre-established groups are broken up. The random group may be comprised of relative strangers and require everyone getting to know each other and initiate respectful behaviour from the start. There can be resistance from students to work in random groups, it is natural to want to work with people they know and have a degree of trust with. Experience in my own teaching has shown that students soon overcome their concerns about working with people they have never worked with before and generally new friendships develop and their network of trusted fellow students grows. It can be argued that to learn to work closely with others that you may not initially know very well is part of experiential learning. This is part of expected professional behaviour. The random approach to forming groups is not perfect but it prevents the clusters of friends in groups or sub-groups. One variant of this to consider was reported by Peter Bier (see the case study in Chapter 4) who used random grouping initially to form groups, but due to the imbalance in gender, he adjusted the groups so there was always two or more female students in a group that had female students [30]. This approach is similar to recommendations in [224].

- If you bring in colleagues or guests, consider that those individuals may be role models too. I have found bringing in women engineers as guests have brought about feedback from students that showed seeing someone like themselves working in engineering was inspiring.

- Prepare the ground work for group or partner work. In a non-capstone project I have found it valuable to spend time at the start of the course to have groups consider what is teamwork, including diversity to ensure that roles and responsibilities to the team are discussed. One exercise for a new group is for them to discuss and write down what they consider a good team member does (within the question set in Table 6.2). This can bring out into the open ideas that all can consider, discuss and support. When I have done this with groups there have been preceding questions about how communication will be done, recordings of decisions will be made and shared, how roles will be respected and how lateness will be dealt with. Again establish ground rules to reduce exclusion, whether intentional or not, or doing the work of another. Such simple rules can be a quick way of reducing problems down the road. Newly formed groups may not consider the organizational framework of how they will proceed. A little help and guidance, along with supporting talk about group work, can help make students aware of their responsibilities to others in their team.

As can be seen from the data there are some similarities between graduates and the general population, but there are groups who are underrepresented in Engineering Schools. Having a diversity issue, where the profession does not have the same distribution as the population, has a compounding effect in that there are fewer role models for students. The lack of diversity of engineering faculty can be addressed in hiring, but still, there can be a wide imbalance in the faculty diversity. Within the family and friends of a student there may be an engineer that can act as a role model and provide encouragement.

The profession itself suffers due to the lack of diversity of its members and perhaps limited perspectives to solutions to engineering problems. Engineering is not the only profession with a diversity issue and it may learn from the success and possible failures of those other professions to open their doors to all. The engineering profession also has a responsibility and ethical requirement when it comes to all. In Canada, like other parts of the world, to be called an 'engineer' you have to be licensed to practice and with that comes legal obligations and requirements. In the code of ethics point 9 is a requirement to the equitable treatment of all, this is summarized by Engineers Canada as "Treat equitably and promote the equitable and dignified treatment of people in accordance with human rights legislation." [49].

How can issues with diversity affect students in experiential learning situations? Conversely can experiential learning impact positively on diversity in the near and longer-term future? These are questions that an instructor should consider before the running of their course that includes experiential learning.

10.3 Indigenous Contributions

If you are a reader from a country which in the past few hundred years has been a colonizing nation rather than a colonized one, then the relevance of this section may not be as obvious to you as a colleague who is from a colonized country. Responsibility to understand the impact of colonialism falls on those with roots in the colonizer. Some may not realize the positions of privilege they have from colonialism and similarly, they may be oblivious to the plight of others. However, awareness is growing on the impact of colonialism. Many countries that have been colonized in the past are grappling with how to deal with the injustices that have arisen out of the colonizing past. Land rights, inequities and racism are all challenges that are at some stage of being examined and attempts are being made to resolve. In 2007 The United Nations passed a Declaration on the Rights of Indigenous Peoples which sets minimum standards for the rights of Indigenous people in 46 articles [194]. Within Article 31 it states:

> "Indigenous peoples have the right to maintain, control, protect and develop their cultural heritage, traditional knowledge and traditional cultural expressions, as well as the manifestations of their sciences, technologies and cultures, including human and genetic resources, seeds, medicines, knowledge of the properties of fauna and flora, oral traditions, literatures, designs, sports and traditional games and visual and performing arts. They also have the right to maintain, control, protect and develop their intellectual property over such cultural heritage, traditional knowledge, and traditional cultural expressions."

As a profession, engineering can be involved with respecting these rights and helping to assist in changing the under representation of Indigenous people in the profession, engineering school faculty and student body. Further, the awareness of engineering's role in colonization and the disproportionate affect Indigenous people suffer due to engineering infrastructure failures [37], is perhaps a starting point for change. Moving forward, approaches like co-design [43] may help with the alignment of engineering projects with community priorities and values.

Before going further, I should declare I am a white male of British origin, who is a settler to, and now a citizen of, Canada and I am writing this on unceded land of the Algonquin Nation. I am not Indigenous and I am reporting here on what I have learned, and I aim to continue to learn. As an immigrant and now citizen of Canada I feel it is required of me to know about the Indigenous Peoples of the continent where I reside, especially following the publishing of the Calls to Action of the Truth and Reconciliation Commission [265] (which is a set of calls to redress the impact of the forced separation of children to Residential Schools, where the horror of the separation was compounded by abuses within the schools). I am not Indigenous and I do not claim authority on Indigenous matters, but I have been involved with Indigenous education initiatives in engineering at the university where I teach, as well as working with a network of others organized through Engineers Canada. I do aim here to report on efforts to increase Indigenous inclusion and awareness in engineering education, but recognize that this may have a bias towards viewing through a Canadian lens. It is worth noting that though the term Indigenous is used through this section of the book, I am not assuming a single group of people and similar traditional knowledge. Indigenous people throughout the world have different cultures, languages, traditions and knowledge. Within Canada there are First Nations, Metis and Inuit which have their own history and culture. Under the single designation of First Nations there are more than 50 Nations with 50 languages [50], some of these Nations may have people and communities residing across the border in the USA, which illustrates how colonization and country borders can split Indigenous peoples.

Table 10.4 gives the percentage of the population who are Indigenous for a sample of countries.

TABLE 10.4
Percentage of population that is Indigenous, for a sample of countries. Sources are International Working Group for Indigenous Affairs (IWGIA) [135] and Statistics Canada (Stats Can) [249].

Country	Percentage of population Indigenous	Data source
Aotearoa/New Zealand	15%	IWGIA
Australia	3%	IWGIA
Bolivia	42%	IWGIA
Cambodia	2-3%	IWGIA
Canada	4.9%	Stats Can
Costa Rica	2%	IWGIA
Greenland	89%	IWGIA
South Africa	1%	IWGIA
USA	2%	IWGIA

Within nations there will be Indigenous engineers, but they may be an under-represented group and there may be a general lack of understanding about Indigenous ways and knowledge within a country's non-Indigenous professional engineers. To increase diversity and the involvement of Indigenous peoples in engineering there can be outreach, encouragement in the recruitment of Indigenous students and their support once they have entered the engineering school. That support can include the need to recognize an Indigenous approach to teaching and passing on knowledge may be different to the current teaching and learning approaches typically used in an engineering school. Within the curriculum there can be the development and inclusion of information for all engineering students to help understand Indigenous knowledge, past and current relationships between Indigenous peoples and awareness for reconciliation. The process of Indigenization and decolonization in a Canadian context has been considered by Gaudry and Lorenz who have noted the spectrum of change involves the stages of *inclusion*, *reconciliation* and *decolonization* [105].

Inclusion can be a starting point for increasing the number of Indigenous peoples within engineering. This can start as initiatives to offer engineering camps for children and increased awareness of the profession and engineering degree programs. Engineering schools can also provide increased recruitment of Indigenous faculty that removes the under-representation, and provides support structures specifically for Indigenous students and the involvement of Elders and Knowledge Keepers.

In the area of science and engineering Indigenous knowledge has been obscured by Western science which has become the standard lens through which science and nature is viewed. However, Indigenous knowledge has existed longer than the Western form of science and has been passed on, often orally, for thousands of years. That knowledge often focuses on nature and sustainability and has been developed through observation and refined over time. It is also characterized by a different view of the connection between humans and nature. Although it may seem surprising to have an parallel form of scientific understanding that developed and is separate to established science, it is perhaps not too surprising on reflection, if one considered how scientific knowledge and its application in structures and technologies around the world appeared before the establishment and mass transmission of Western scientific knowledge. For example, consider ship engineering and navigation by seafaring peoples around the world, as well as buildings such as the Pyramids, castles, temples and other structures that remain today. They certainly show detailed knowledge of science, architecture and engineering. So the transferal of scientific knowledge for application and use can be done without the semiotics of Western modern science. Although Western science has become highly efficient in its processes of testing, validation, documenting and transferal, it has also led to technological and environmental disasters that impact people regardless of nationality and origin. Recognizing and learning from Indigenous knowledge offers complementary forms of insight and learning. For example, an Indigenous relationship with nature can be a symbiotic one and more holistic than a Western view. Snively and Corsiglia have detailed some aspects of Indigenous knowledge in science and its implications in science education [245].

Examples of Indigenous engineering and approaches to creating awareness of Indigenous engineering are featured in the ASEE Prism magazine cover story 'Wisdom of the Ages' by Jennifer Pocock [213]. These include mention of living bridges of the Khasi people within India and Ma'dan of the wetlands of Iraq, who suffered under the regime of Saddam Hussein. The article also outlines the pioneering work of Deanna Burgart, a Dene and Cree engineer and faculty member at the University of Calgary, who created IndigeSTEAM and coined the term Indigeneering [213]. Further information on Indigenous engineers can be found from the study of Jordan et al, which looked at 20 Navajo engineers and the relationship of their engineering work and how it relates to their community and culture [143]. This is a valuable study as it provides insight into how Indigenous engineers bring together their Indigenous, in this case Navajo, culture and knowledge with their engineering design and practice. The study took a phenomenographic approach (through interviews) and looked at 20 engineers with experience in practice ranging between 3 to 30+ years. With the area of engineering including software, energy, mechanical, civil, mining and civil. The results found four areas of how the engineers linked their design and practice work to their culture and community. These were identified as: i) Navajo-centred behaviour, ii) Navajo-centred purpose, iii) Navajo-centred strategy and iv) Navajo-centred application. These categories are listed in a hierarchy such that iii) involves i) and ii). All 20 engineers were linked to i) but only five were associated with iv) [143]. The report also raises some of the tension between the Navajo culture and engineering, including a feeling of detachment from their culture. This included challenges to living away from their community when entering college. This was reduced by finding other Navajo students and the American Indian Science and Engineering Society (AISES).

The support for Indigenous students is important to assist the growth of Indigenous student population and their success, whether through university initiatives, including Indigenous centres on campus, or from external organizations like AISES. One example of strong support can be found in the Faculty of Engineering at the University of Auckland, Aotearoa/New Zealand, where an academic and mentoring program called the Tuākana Engineering Program [272] to help Māori and Pasifika students (note there are two distinct Indigenous populations in Aotearoa/New Zealand). The program provides tutorial support specifically for Indigenous students within the core first year and space for this support is provided too. Coupled with a supported recruitment program of students from the Indigenous groups [19]. This current program shows how a holistic approach for Indigenous groups can be developed; from recruitment, academic advising support, student mentoring (from being a mentee to becoming a mentor) and space for these support mechanism to exist. As well there is use of the Māori language, as shown in the university's website. This range of support mechanisms shows how one university is developing a open and supportive environment for Indigenous students, which could overcome some of the feeling of detachment and isolation that the previously mentioned Navajo engineers described.

Support for Indigenous culture can be made within the academic environment. Western science and ideas are the basis for most, if not all, engineering courses in the majority of engineering schools. Providing space for the inclusion of some Indigenous knowledge provides some way of showing Indigenous students their knowledge is important to consider, as well as informing non-Indigenous students about the knowledge and culture of Indigenous peoples. The learning can go further as Indigenous approaches in engineering designs can have broader consideration and balanced approaches, including the impacts on the physical, societal and emotional level. This knowledge can be beneficial to non-Indigenous students. This learning has to be done appropriately and with respect. Many Indigenous centres of campus can help and advise here. Here is an opportunity to include an elder or a knowledge keeper who can talk about the knowledge. The benefit is that graduating engineering students, who are not Indigenous, can gain more insight and appreciation so when practicing as engineers if their work involves working with Indigenous peoples and their land then there can be a deeper understanding. In Canada there is a growing understanding of the need for this and many engineering schools are developing approaches to include more Indigenous knowledge, especially as the buildings and campuses are on unceded First Nations or Treaty land [235, 236]. Indeed the recognition of Canadian universities being on Indigenous land is being made regularly at events, such as meetings and convocations, and even in email signatures. This acknowledgement becomes a respectful reminder of the past, which has led to situations today.

One way forward to include Indigenous knowledge into the curriculum is the approach of Two-Eyed Seeing. Developed by Elders Marshall (of the Mi'kmaq Nation) and Prof Cheryl Bartlett at Cape Breton University, for use in Integrative Science [25]. The fundamental idea with Two-Eyed seeing is "learning to see from one eye with the strengths of Indigenous ways of knowing and from the other eye with the strengths of Western ways of knowing and to using both of these eyes together" [119]. As Hatcher et al explains the approach of Western science, commonly know as science, is based around an impartial and separated analysis of natures, often split into areas or disciplines starting with distinct fields such as physics, chemistry and biology and subcategories can grow from there [119]. Alternatively, Indigenous science may not separate the individual. There can also be an acknowledgement that there is an inter-connectedness to things and this can build in a sense of kinship and responsibility. Of course, each Indigenous peoples may have their own local knowledge and context, so I accept there are generalizations here, in an effort to show the different epistemologies that the concept of Two-Eyed Seeing tries to help to resolve. This concept has been used beyond North America [184, 204] and is one attempt to resolve the Western and Indigenous ways of knowing and the role that Indigenous knowledges have in helping all [186].

As Indigenous science is often land-based it can be rooted in participation within the environment and the relationship to the environment. The Intergovernmental Platform on Biodiversity and Ecosystems Services (IPBES) adopted a policy that recognizes and incorporates Indigenous and local knowledge (ILK) in 2017 [125]. They define local as follows "Local communities are groups of people who maintain inter-generational connection to place and nature through livelihood, cultural identity, worldviews, institutions and ecological knowledge." Which includes more than Indigenous groups. There is recognition that ILK is holistic and can involve naming and classification, governance, economics and spirituality, amongst others. Correspondingly ILK can be diverse due to geography and history of interaction of the land and the culture that has developed the knowledge can have its own way of validating and sharing the knowledge. Overall there is a recognition that knowledge extends beyond the traditional Western-based scientific knowledge and that there needs to be a recognition, space to include and respect for ILK.

The term Traditional Ecological Knowledge (TEK) is a term used to describe the Indigenous understanding, science and knowledge and could be an alternative to ILK. The focus of being land-based draws in the label 'ecological' and the term 'traditional' may imply a static and unchanging knowledge that is passed along. Although it is passed on this knowledge is not static, but evolving by the users [211]. The knowledge may involve cultural beliefs and practices and may be orally passed on [192]. Whereas Western views and knowledge of the nature can separate humans from the natural world and at the extremes lead to two different approaches to how to interact with nature. One being the use of resources and its extraction and exploitation, and industrialized approach to nature. The other view is to leave the natural world untouched and to protect it, the parks and wilderness approach to nature. Pierotti and Wildcat argue that TEK provides a third approach to interacting with nature. Humans are an implicit part of the natural world, not separate from it and there can be a careful balance between extraction and conservation [211]. Significant care needs to be taken in considering and resolving TEK and Western science. One takes a holistic approach and the other attempts to split and categorize into sub-topics and specialization. A valuable source of information on the challenge of combining TEK and Western science is Nasdasdy paper that came from 32 months of anthropological fieldwork in a community in the Yukon Territory, Canada [192]. The challenges of resolving Western science with TEK are highlighted, and discussion of the compartmentalization and distillation of TEK from its holistic, including social context is made. Within the paper is a revealing example on how a combined group of scientists and First Nations community members jointly did a counting survey of sheep. Working together the total count was made and agreed upon. The problem then comes with interpretation. The paper explains that the scientists figured the low count was too low of a sample size, whereas the members of community involved with the count considered the low number to be significant and showing the decline of the sheep population [192]. Faced with a jointly made study and agreed upon results, the two forms of knowledge brought about different conclusions. The previously mentioned idea of Two-Eyed seeing may be one way of helping resolve the difference in understanding between Western and Indigenous knowledges.

Before leaving this section on learning with Indigenous peoples and groups it is important to mention that non-Indigenous people involved in such collaborative experiential learning need to understand more about the Indigenous group and to have knowledge about allyship before undertaking close work. Many Indigenous groups have suffered loss due to colonialism. That loss could be land, language, culture, knowledge and members of their community past and present. It would be incumbent to anyone bringing non-Indigenous students to work with Indigenous communities to prepare the students with a fundamental level of information to help them with the understanding the background to the Indigenous collaborators. Allyship requires self-education about Indigenous matters and should not rely on the Indigenous peoples having to fill in the basic knowledge gaps. Most non-Indigenous people have benefited from colonialism (universities have roots in colonial systems) and that benefit needs to be recognized. The building and developing of relationships with Indigenous peoples is important and a fundamental part of allyship. Providing students with basic knowledge can not only help initiate relationship building, but will also help understanding in the future. Here experiential learning can not only help with technical understanding but it could draw in historical, political and social context too.

10.4 Ethics

Ethics is a part of the profession of engineering. With engineers working on complex problems that can involve innovation and construction of items that can have social, environmental and economic impact, then this can raise moral and ethical questions with respect to the design and construction in engineering [75]. In some jurisdictions registration as an engineer can have legal requirements and responsibilities. For example, in the Canadian province of Ontario the legislation and laws around the practice of engineering are detailed in the Professional Engineers Act.

Ethics is usually a part of the engineering curriculum and can be included in either a separate course, perhaps on professional practice, or included in parts in different courses. The range of content in ethics can be wide, however, from a focus of the 'rules' relating to engineering practice (conflicts of interest for example) to more open areas such as engineering and relating to the sustainability, humanitarianism and design aspects.

As many engineered systems grow in complexity there can be ethical challenges faced by engineers now and in the future. Ethical concerns that may not be so obvious at the time of design. For example, consider the possible personal data collection through mobile phones that could intercept personal information, such as location or health and financial data. Data that may be valuable to the phones owner (hence the use of a smartphone), but potentially open to misuse by others. The Volkswagen emission scandal is another example, where sensor information in a diesel car could be used to detect when the vehicle was undergoing an emissions test. Then the vehicle could make adjustments to reduce the emissions to be under normal operation emissions [12].

The challenges and potential misuse of such systems (and large amounts of data) is being raised more often and a potential point of concern. With the current and rapid development of more sophisticated systems, such as autonomous vehicles, robotic systems and smart cities, there is an increasing need for engineers and student engineers to be aware of potential ethical issues [88, 136, 149]. This goes beyond the more traditional ethics codes and laws (such as applying equity in hiring and clear financial record keeping and reporting).

Experiential learning may provide opportunities to consider the ethical challenges when working as an engineer. In a design project, or on a co-op position, there is perhaps more of a context to see how ethically challenging situations could arise. There is then the potential of incorporating ethics education within existing experiential learning, or developing ethics in new courses from the start. In this section we will consider some of the literature relating to ethics and consider the needs and approaches to be taken.

Haws provided one of the early 21st century examinations of ethics in engineering education [120], having examined conference proceedings from 1996 to 1999. The study showed six ways that ethics was approached in education. These were [120]:

- Professional codes
- Humanist readings
- Theoretical grounding
- Ethical problem solving heuristics
- Case studies
- Service learning

The last two areas fall into the area of experiential learning, especially the last. The first four may more suited to class instruction and discussion. Outside of the first one, professional codes, perhaps many engineering instructors will feel poorly qualified to teach humanist readings, theoretical groundings and ethical heuristics. Haws identified that some non-engineering instructors were teaching the ethics [120]. The paper highlights that effective teaching of ethics should go beyond the codes of ethics (identified as one of the listed approaches) that most national or regional regulatory engineering bodies may have in place. These codes typically list that public safety is paramount, practice within your expertise and working with integrity for clients and employers, amongst others. Haws points out that

> "These [codes of ethics] are not appropriate terminal levels of development for engineers. If they relied on as such, it will restrict our students' ability to reason through their own values, and select ethically appropriate courses of action." [120], p. 224.

He also notes that the code of ethics an engineer may work under is "more important to our student's professional development than to their ethical development." [120], p. 224.

So if these sets of rules or guidelines are a starting point of ethical education, how can an engineering student's ethical awareness and education be extended beyond them, so that they can deal with more nuanced ethical situations?

The most popular approach of the above list was the case studies. There have been many high profile engineering case in recent decades. Haws points out a number including the Challenger disaster, Chernobyl and Exxon Valdez. More recent cases that could be the previously mentioned Volkswagen emissions scandal [12] and the Boeing 737 Max crashes [106]. These high profile cases may show the impact around ethical decision-making, though they may be viewed as extreme (most are disaster focused) and some students could consider them beyond their likely experience in work. Potential use of other smaller scale case studies could make the case studies more relatable. Haws does highlight that if case studies are used without theoretical underpinning of the ethics then there can be a range of perspectives given which could lead to ethical relativity.

Of the listed pedagogies Haws considers that the service-learning approach offers significant potential for seeing engineering impacts through another's perspective. This can lead to divergent thinking, which if combined with theoretical ethics underpinning can lead to effective education.

In 2018 Hess and Fore conducted another survey of US engineering ethics literature to look at the strategies for teaching ethics in engineering programs [123]. Since Haws' 2001 work there had been an increasing of the emphasis on engineering ethics following revisions of the US accreditation (ABET) criteria. So in the time between the two surveys there could be information in the literature on the strategies of how different schools and programs had included and adapted professional ethics teaching and pedagogy. The survey process resulted in examining 26 different articles [123]. Again, a significant number, 21 of the 26, were case-based studies. Though the most popular pedagogic category, 22 of 26, was codes or rules in ethics. There were 11 of the 26 that involved philosophical ethics and 12 of the 26 that looked at ethical heuristics. Some approaches used debate or discussion, 20 of the 26. Less common approaches were peer mentoring (3 of 26), micro-insertion (2 of 26), game-based pedagogy (2 of 26), community engagement (2 of 26) and real-world exposure (2 of 26). We can see from some of these areas that reported approaches fit within the realm of experiential learning, whether community engagement, real-world exposure or case studies. Hess and Fore focus on four different exemplars within their paper and the micro-insertion is one example studied [123]. This too could be used in a classroom-based experiential learning as the example paper by Davis [71] takes a typical engineering problem and with some small (or 'micro') adjustments shows that there can be ethical questions arising from an engineering calculation. For example, a calculation of performance can then be overlaid with issues of the hazards of the materials that provide that performance. Davis' micro-insertion shows how a small change can raise an ethical issue and consideration in a regular technical course calculation. There is some alignment as to how ethical situations can occur in the engineering profession. It is unlikely that an ethical issue will come clearly identified, unlike a case study in an ethics class. The real-world ethical challenge may not at first be detectable. The micro-insertion approach is effective in that it shows that calculating or determining the 'best' solution may have impact on final proposed solution. The student starts to learn that the final solution to a design issue can come from understanding safety concerns and requirements as well as other factors like cost and availability.

Ethics and ethical issues are perhaps not on the forefront of students minds as they undertake their engineering education. This is a topic that has to be introduced and shown to be relevant to their studies and future careers. As has been noted, the blunt use of a set of rules only starts the conversation. There are more subtle levels that can be included, especially as technology becomes such an integral part of everyday living. The use of case studies allows discussions in courses that are not primarily designed to be about ethics. I have introduced case studies in a third level (non-capstone) project course. As students were focused on the design on their project it seemed an opportune time to discuss ethical issues in engineering. In part it was to prompt students to zoom out from the detail of their design focus and to encourage them to consider engineering projects in a broader way. With students having encountered ethics in their first year courses and again in their fourth and final year, it also seemed useful to reconnect to the topic to recall and prepare. In a similar way the micro-insertions could be a planned way of included a conversation on ethics in a problem from a technical topic course. Ethics occurs in engineering in a manner that is not clearly designated, so having courses connect back to ethics seems valuable.

For experiential learning outside of projects, service learning or active classrooms there is likely to be less controlled opportunities to raise ethics. That is not to say that it may not occur but hopefully a supervising engineer on a co-op term will help identify, educate or resolve any ethical issue. Laboratory work can deal with the incidents of modified or fabricated data, should a student makes a bad decision when conducting a laboratory exercise. Any academic integrity failure by a student is an opportunity to reinforce the ethical behaviour of an engineer. Having been an associate dean and had to deal with academic integrity cases I would try and make any lapse in academic integrity an opportunity to educate. The unethical short cut they tried to take, or the lapse of judgement, then provides a way for the student to reflect and learn. Hopefully, there are few students that need to learn this way.

For the more theoretical and deeper examination of ethics and moral issues, there may need to be the use of ethicists or philosophers. Again in my project course, I have had ethicists as guest speakers. To have an expert discuss the ethical dilemmas associated with autonomous vehicles, for example, can show how complex some of the issues are and how seriously they are being taken. It provides a different perspective for the students and helps them consider the broader issues in engineering and the ethics surrounding them.

10.5 Chapter Summary

In this chapter we have taken a step away from looking directly at forms of experiential learning and considered some of the social issues around engineering. These can not only affect experiential learning, but the can also be considered when deploying that type of learning. The areas considered are one of just a few of a range of possible areas. We could have considered sustainability and climate change. The topics selected have been chosen to show some of the current challenges that provide ongoing dilemmas. The lack of diversity in the engineering profession has been a long ongoing challenge which seems to be slow to change. Awareness and understanding can help both change to more of an equitable balance in who works as an engineer. The Indigenous section is an attempt to show how this area is more than just a diversity issue in engineering. If you are living in a country with no designated Indigenous people then I hope you do not skip that section. As engineering work becomes more global and projects can be multi-national, appreciation is needed that some of our students will work on projects with Indigenous engineers, peoples and their land. It is also worth considering the impact colonialism had and still has on populations around the world. Engineering was used in the process of colonialism. We can see also that Indigenous section is somewhat of a bridge between the diversity and the ethics sections. Each of these sections moves beyond just the technical and have been included to show how factors can affect experiential learning and how this type of learning can help educate future engineers for the challenges in the social arena.

11

Final Pieces and Conclusion

11.1 Introduction

In this final chapter we will look briefly at some remaining topics and then draw some conclusions on the central theme. The chapters in this book have looked at the various types of engineering education where experiential learning can occur. I have taken a broad view of where experiential learning can occur. This includes the classroom, where it may not be expected to be found. Some may consider experiential learning to be restricted to co-op and project work. But, as we have examined, the ways of students learning by doing can go further than just those more obvious forms.

There some topics which I would like to deal with in this final chapter. These include assessment and accreditation. Within various chapters assessment has been mentioned, but since it is integral to evaluating a student it may be worth revisiting again. Similarly, I have not discussed accreditation. This too can be a significant force in engineering education. It can have an impact on experiential learning and as most accreditation assessment is outcomes focused, experiential learning can provide good evidence for accreditation assessment.

It will also be useful to return to looking at the models of learning and having one final consideration of experiential learning through the lens of the learning models. It may also help to consider their value too.

Lastly, we will look a little towards the future and consider some of the possible changes that can occur in the curriculum and where experiential learning can help.

11.2 Assessment

Many classroom-based engineering courses will most likely have a scheme for assessing students-based around an examination, with possibly mid-term tests,

DOI: 10.1201/9781003007159-11

related lab reports and assignments included. Most, if not all, of the assessed work will be done individually. With experiential learning-based courses, as we have seen, the approach to the learning experience is much more immersive to the student and the assessment needs to match the authenticity of the learning experience. Put another way, it would seem unusual (but not impossible) to apply a written exam to say a capstone project. It would more appropriate in this example to have reports, such as interim design reports or a final report, and an oral presentation. Indeed, the oral presentations could be in design reviews with questions asked by the attendees, or in a thesis defence format. The authenticity of the challenge of a capstone project then follows through into the assessment, making that also become part of the learning and consequently a type of formative assessment, though the reports and other final assessed work can become summative too. The categorizing into formative and summative can be blurry and debatable, but the benefit of experiencing the production and presenting a poster of some undergraduate research work, will very likely be done with instructor/supervisor feedback on drafts. So, we can see that the learning can go right through into the assessment process of experiential learning.

If there is not an examination, what then are the forms of assessment of experiential learning? Possible types are shown in Figure 11.1, split into

	Exam / Test	Problem / Calculation	Essay	Report	Presentation	Reflection	Literature Survey
In–class	●	●	●		●		●
Laboratories		●		●			
Projects				●	●	●	●
Co–op				●		●	
PBL		●			●		●

FIGURE 11.1
Matrix of assessment types for different forms of learning. This may not include all assessment types but is intended to cover the common forms. These assessments produce some form of output, commonly written. However, for *projects* and *co-op* there may also be some form of subjective performance assessment of the student by a supervisor.

categories matching some of the chapters of this book. It can be seen that reports can be used in a variety of forms of experiential learning. Though they can be at early (proposals) or mid (interim reports, reflections) stages of the course, the main reporting stage is at the end of the period of learning. This final summing up of work, done into a cohesive report, with background and summary sections becomes an important piece of work for a student to undertake. It should draw the project together and it can help the student process the overall work, information and experience. If the project has been done in groups then it is likely the report is split into sections. This can make a comprehensive document on the project and help in experiencing the collaboration with project colleagues to produce a coherent document. Again, another vital learning experience at the assessment stage.

One challenge with reports is they are commonly at the final stage of the project. This can mean there is little or no opportunity to give feedback on performance and achievement (supervision may provide some indication of how progress on the task is going). So if there is to be earlier assessment to provide feedback there needs to be some appropriate mechanisms. For reports and longer activities, like a one or two-term report, there can be a proposal document, interim reports, design reports and reflections. Proposals are effective to focus the students on what they need to do and how they will go about the work. This can include planning, Gantt charts, initial specifications and other preliminary analysis. If the project is a two-term project or longer there can be a call for an interim report after the first term. That can show what progress has been done on the project and what needs to be done. For a co-op placement this can help in checking how the student is progressing and if there needs to be further assistance and guidance needed (again, regular supervisory meetings should reveal that though).

Design reports can become focus studies on a particular aspect of the design. In a group project this can be an opportunity to have an individual report. A student within the group will have worked on a particular aspect of the project and the is the opportunity to have the student write about that aspect and to detail the design and performance. If the project is to continue beyond the currently registered group of students, those design reports can become helpful to the next batch of students who take the project further. As each topic should be individual it is possible to have other students as second readers and checkers of the report. There is then a passing on of knowledge, a chance for students to see their colleagues' work and to help check the clarity an content of the report. From a situated learning perspective there is a sharing of knowledge between the students, as well as taking a more professional approach to group tasks. I have been involved with projects where student peer checking is part of the reporting process and it helps with exposing the students to different, but connected work. This allows them to see the calibre of other's work and to take some responsibility of checking for errors in the report. In these projects the checker's name is also included on the reports and it is emphasized that they take some responsibility if mistakes or omissions get through to the project supervisors. There is a secondary benefit that there has to be an earlier deadline for the checkers to be able to do their reading.

Reflections can be an interesting form of assessment. This can be a way to encourage students to think about the work or topic they are reflecting upon. It can be a way that the various challenges a student encounters can be brought to light and considered openly by the student, in a way that a typical engineering report cannot do. Challenges with understanding a particular problem, whether it is in construction, project management and interpersonal issues can be discussed in a reflection. Having used reflections for a number of years in a third-year project course I have found that I can get to know about some of the challenges and issues that may not be readily on display within a laboratory room. As well, you can read about how students are solving problems and what they are learning. There can be a strong benefit of having a student do a reflection. It develops self-learning and criticality in how one approaches work. It can be a way to encourage a student to think about a particular aspect and to delve further with their thoughts and consideration. The reflection can become a personal dialogue. Having the reflection as a graded assessment is a way of encouraging it to be done and for there to be a grade benefit for doing it.

Sometimes students can find the reflection assessment a challenge. I have found providing a potential reflection topic to be valuable. It gives a suggestion to help initiate a reflection. Ask students to consider how they approach their work; for example, what steps do they take towards solving a particular problem. Or, asking them if they could restart a task, how would they change their initial approach and why. These prompts can help students, who may not have written many reflections before.

Sometimes I find some students have difficulty in moving beyond a basic description of what had been done in the project over a particular period, rather than providing a reflection. Encouraging the students to think further and move them towards deeper reflection and perhaps even towards more metacognitive approaches to their work can be a challenge. However, it is worth at least starting that process. I have used a rubric approach to the assessing of a reflection, Table 11.1. This can give a guide to what is expected on their reflection. (I also assess their written communication but I do not give that rubric here.) The rubric guides the student to consider the analysis of an experience and to embark on considering outcomes. The reflections are done individually and I mention in documentation who may see their reflection, such as possibly a TA, or an accreditation visitor. As a course leader I have found it valuable to know what the students are considering and reflecting on and I personally find the written individual reflections insightful to see what has been challenging and what they are learning from the course.

For group-based reflections I have given to students an oral presentation-based reflection of "One thing learned from the project". This causes the group members to have to discuss amongst themselves as what they believe they have learned and what they think is important. The results can be interesting

TABLE 11.1
Reflection rubric used with project reflection.

Level	Expectation
1. Beginning	Only a description of experience. No significant reflection given.
2. Developing	Experience described and related to other experiences, ideas and/ or resources. However, little analysis and/or decided outcomes.
3. Accomplished	Experience(s) well described with a thoughtful analysis. Concluding ideas, outcomes or resulting actions described.
4. Exemplary	Critical analysis of described experience. Well considered details of outcomes and actions given. Extended conclusions for further experimentation and relating to further ideas.

with non-technical topics occurring, such as the importance of communication within a team. By making them give brief presentations to other groups the other teams of students can learn what each group has considered to be important. Making the students in the audience consider their group's response to the other groups'. Again, it is valuable for the instructor to see and hear what students consider has been their key learning experience.

In a study looking at assessment types in experiential learning in community service learning Chan has looked at forms of assessment in the literature, and also reports on a study undertaken by them with engineering student's working on a community service project [61]. The assessment approaches highlighted in the literature includes direct observation, presentations, oral assessment through questioning and reflection journals. The project described in the study used a daily reflection journal, amongst other forms of assessment including written pre-trip expectations, daily performance, team building performance and a final report. Feedback from the participants indicated that maintaining a daily reflection journal was challenging, especially after the daily work. This raises a point about the frequency of reflection and this may be something for an instructor to consider if they want to use reflections. Not only from a workload perspective but from the point of view of how students need time to process what they are learning from the experience. People who have looked at reflections may note how there may be a irregular pattern to insights to the learning. When some experience and understanding come together there can be a reflection that recounts an 'a-ha!' moment of insight, before returning to more regular and lower level insight. Reflection needs time, but also there needs to be some frequency, otherwise it can become a single looking back reflection at the end of a term (which still has merit as an exercise). Perhaps weekly or monthly is the frequency of merit. Chan also raises the point of exploring the use of media beyond the written word, such as photographs, audio recording and videos [61]. This is an interesting area to explore, especially at a time now when most students carry a camera and video/audio recorder with them, in the form of a smartphone. In a third year group project course I have started to include video presentations at the end of the build. This allows students to show the operation of their project (in this case an electronic project) and to guide the viewer through the detail. Over the duration of the course I have students give presentations to all groups so that the whole class can see the evolution of the project, from proposal to completion.

When it comes to reflections it can be worth considering how deep, surface and strategic learners can approach reflecting about the course. Young conducted an examination of reflection on deep and surface learning of 214 students in a marketing analysis course [290]. Students were involved in a reflection intervention, including a take home exam question on reflection, within the course that used a flipped classroom approach and was guided by the Kolb Experiential Learning Cycle [290]. An online survey of students at the end of the term was used to collect information on deep learning. The author reports the study showed support for the idea that reflection and deep learning motivation are linked, as well as "reflection fosters deep learning while discouraging surface learning" [290]. This indicates that a reflection can help develop deeper learning, so this may move strategic and even surface learners towards deeper learning. Providing guidance on the reflection process may be required, as well as stressing the importance to the professional need of undertaking reflection on their work and approaches to work.

Other types of assessment such as literature surveys, papers on research or analysis, posters and oral presentations, can mimic the form of work in research or industry. These can help prepare the student for future graduate studies or professional work. This relates back to the earlier mentioned authentic nature of assessment in experiential learning. Presenting a paper or poster in an undergraduate conference, design review or course final showcase can help develop those communication skills, as well as provide assessment. From a situative learning framework this type of output can be encouraging to the student as they may perceive that the are doing more of what could be encountered in the profession. Presenting a paper at an undergraduate, or a professional, conference can be used as a final deliverable in an undergraduate research course. Feedback on these types of assessment can also help them develop the techniques in producing the respective form of output, produce confidence and so better prepare the student for the future.

This brings us to what the student perception is to the forms of assessment used in experiential learning. Wilson, Yates and Purton at the University of Saskatchewan have looked at this, along with performance and preference [282]. They investigated these aspects using a presentation, case study, essay and journal. The most preferred was the case study and presentation, which were group based and collaborative. The study's students thought they helped in a number ways, including being engaging, challenging, helpful and beneficial. The least engaging was the journal, which was done individually, but the authors reported this showed the most understanding. The essay, being perhaps the most familiar form of assessment to the students, was valued for its structure (familiarity may be a factor here) but not preferred over the case study. What is interesting here is the preference to the group/collaborative work, though the individual journal showed better understanding despite it being considered to be not engaging. The mix of perception and benefit is not as clear as maybe thought at first consideration.

As we approach the end of this section it is worth including a final quote from Wilson, Yates and Purton's paper that says,

> "We recommend that instructors using EL methods strongly reflect on the alignment between teaching and assessment methods and look for opportunities to teach and assess collaboratively." [282]

11.3 Accreditation

One area that has not been significantly discussed so far is accreditation and how this links to experiential learning in engineering education. Because of the professional status of engineers there is the oversight of the undergraduate education of engineers in many countries. This maintains a level of quality of education and expectation on a graduate's ability after completing their studies and then embarking as an engineer in training (EIT). After suitable experience as an EIT and fulfilling any other requirements, the individual can apply for membership to the appropriate jurisdiction's licensing body. For some countries, you cannot call yourself an engineer without being licensed. As engineers work with others in different countries, it was recognized there was a need for some commonality in the standards of engineering education, along with expectations of competence of an engineering graduate around the world. This assurance of standards could then help aid the mobility of engineers. In 1989 six countries signed the Washington Accord, Australia, Canada, Ireland, New Zealand, the United Kingdom and the United States of America [6]. The Accord recognized the equivalency of each nation's accreditation of engineering programs, allowing graduates from different signatory countries the rights and privileges in another country or jurisdiction and it also allowed the sharing of methods and approaches to accreditation [6]. As of the writing of this book there are now 21 signatories, see Table 11.2 for the list of countries and the overseeing organization.

Over time, and as the number of signatories has grown, the Washington Accord signatories have moved from an informal agreement to formalized arrangements relating to accreditation and in the 2000s there was a move towards outcomes-based assessment of competency and the use of graduate attributes. This route was not intended to be a prescriptive approach but

TABLE 11.2
Signatory countries to the Washington Accord [9].

Country	Representative Organization	Year
Australia	Engineers Australia	1989
Canada	Engineers Canada	1989
China	China Association of Science and Technology	2016
China Taipei	Institute of Engineering Education Taiwan	2007
Costa Rica	Colegio Federado de Ingenieros y de Arquitectos de Costa Rica	2020
Hong Kong China	Hong Kong Institution of Engineers	1995
India	National Board of Accreditation	2014
Ireland	Engineers Ireland	1989
Japan	Japan Accreditation Board for Engineering Education	2005
Korea	Accreditation Board for Engineering Education of Korea	2007
Malaysia	Board of Engineers Malaysia	2009
New Zealand	Engineering New Zealand	1989
Pakistan	Pakistan Engineering Council	2017
Peru	Instituto de Calidad y Acreditacion de Programas de Computacion, Ingenieria y Tecnologia	2018
Russia	Association of Engineering Education of Russia	2012
Singapore	Institution of Engineers Singapore	2006
Sri Lanka	Institution of Engineers Sri Lanka	2014
Turkey	Association for Evaluation and Accreditation of Engineering Programs	2011
United Kingdom	Engineering Council United Kingdom	1989
United States	Accreditation Board of Engineering and Technology	1989

rather a way that signatory organizations could show equivalency through the common outcomes [7]. The current graduate attributes are listed in Table 11.3.

It should be noted that there may be other graduate attributes at an academic institution that is required for academic quality assurance purposes of the country, region or state. There may be similarities between the graduate attributes of the quality assessment and those of the Washington Accord, and there may also be differences. We will restrict discussion to the Accord.

These Washington Accord graduate attributes draw from their knowledge and attitude profile which is given in Table 11.4.

TABLE 11.3
Graduate attributes [8].
* Previously split into two categories, 'Engineer and Society' and 'Environment and Sustainability' and merged in June 2021.

	Graduate attributes
1	Engineering knowledge
2	Problem analysis
3	Design/development of solutions
4	Investigation
5	Tool usage
6	The engineer and the world*
7	Ethics
8	Individual and collaborative teamwork
9	Communication
10	Project management and finance
11	Life-long learning

Seeing the graduate attributes in Table 11.3 we can see that experiential learning can help considerably to collect evidence that students have attained levels of competency in the attributes. Let us consider three different activities in courses.

1. Undertaking a group project in a full-term project course can involve the following: Working in a team can involve Individual and Teamwork. Scoping the project could result in Problem Analysis. The prototyping and construction of a solution can involve Design and Development of Solutions. The writing of a report and giving of a presentation can involve Communication. The organization of time and resources can involve Project Management and Finance.

2. A class-based problem done in a group of three students can involve: Problem analysis in undertaking the challenge. Individual and Teamwork as the group works together on the problem. Communication of the results to the class.

3. An individual student's research-based project on an advanced engineering topic could involve: Engineering Knowledge acquisition when learning about the topic. Investigation when looking at advanced texts and papers. Problem Analysis with looking at the research question. Modern Tools when perhaps using laboratory equipment or simulation tools. Techniques that can be applied to Life-long learning is being used and understood.

In each of the three examples it can be seen to be there are ways that the competency of the student can be assessed, along with evidence being collected. The instructor can see how the student is matching the expectations

TABLE 11.4
Washington accord knowledge and attitude profiles for engineering graduates [8].

Item	Knowledge and attitude profile
WK1	A systematic, theory-based understanding of the natural sciences applicable to the discipline and awareness of relevant social sciences
WK2	Conceptually based mathematics, numerical analysis, data analysis, statistics and formal aspects of computer and information science to support detailed analysis and modelling applicable to the discipline
WK3	A systematic, theory-based formulation of engineering fundamentals required in the engineering discipline
WK4	Engineering specialist knowledge that provides theoretical frameworks and bodies of knowledge for the accepted practice areas in the engineering discipline; much is at the forefront of the discipline.
WK5	Knowledge, including efficient resource use, environmental impacts, whole-life cost, re-use of resources, net zero carbon, and similar concepts, that supports engineering design and operations in a practice area
WK6	Knowledge of engineering practice (technology) in the practice areas in the engineering discipline
WK7	Knowledge of the role of engineering in society and identified issues in engineering practice in the discipline, such as the professional responsibility of an engineer to public safety and sustainable development
WK8	Engagement with selected knowledge in the current research literature of the discipline, awareness of the power of critical thinking and creative approaches to evaluate emerging issues
WK9	Ethics, inclusive behaviour and conduct. Knowledge of professional ethics, responsibilities, and norms of engineering practice. Awareness of the need for diversity by reason of ethnicity, gender, age, physical ability, etc., with mutual understanding and respect, and of inclusive attitudes

and if there is need to change the learning. That evidence that is collected can also be provided to accreditors at the time of an accreditation visit. Collecting evidence for some of the graduate attributes may be challenging. For example, Individual and Teamwork can focus on productivity and output, along with observation. But sometimes there are interactions that occur out of sight of the instructor and it may be important to capture those. Using a situated model it can be those interactions that can facilitate learning. With a project course with multiple teams I have used a group evaluation form, Figure 11.2, and

Group Self-Evaluation

Group (number and letter)

Group Members and Roles

Project Title

Please answer the following questions by circling the appropriate level.
For this course ...

Did you achieve your various group deadlines?	Often missed them	A few were missed	No important deadlines were missed	No deadlines were missed
Did you achieve your original technical goals?	Missed most	A few key goals were missed	No important goals were missed	All goals were reached
Rate your overall group attendance in the lab.	Someone was late, left early or absent every session	Someone was late, left early or absent each week	Occasional lateness, early leaving and absence once a month.	Only one or two shortened attendances or absences over the term
Did differences of opinion and any conflict within the group get resolved?	Rarely	Occasionally	Most of the time	Always
How would you rate the trust and dynamics within the group?	Little trust and poor interactions	Some trust and minor conflicts	Dependable team members and mostly good interactions	Very dependable team members and very good interactions
Overall rate your group performance	Poor	Developing	Accomplished	Exemplary

FIGURE 11.2
A project group self evaluation form to examine team performance and dynamics. This is completed in consultation with a supervising professor, allowing for any clarifications.

had discussions with the group whilst completing it. This can act as a form of concluding feedback and reflection for the project team. After completing that I have also provided some individual evaluation forms for completion in private. This includes the question:

> "In your opinion who was the group member that provided you and/or the group with significant input on technical matters, and/or who showed leadership (does not need to be a designated group leader). You can name none or more than one. Briefly explain why."

This I have found to be a useful way to determine who has been providing help and guidance within the team. This serves as an illustration that collecting information for the graduate attributes can help in the evaluation of students as well as help with providing insight and encouraging reflection for the students and the instructors.

11.4 Learning Models

In this final chapter it is worth returning to the learning models and in particular situative learning. Throughout this book there has been numerous reference to this model as I believe this model fits well with experiential learning. My opinion on this was reinforced by the COVID-19 pandemic and the drastic changes to delivering courses. I had to adapt a third-year electronics project course to being delivered online. Moving a more traditional course online was one level of challenge, but moving a laboratory or even a project course online poses another level. Laboratories comprise of dedicate space and equipment for undertaking technical work and it is a challenge to duplicate this at a student's residences. Fortunately, the electronics project often used a microcontroller like an Arduino, or a microprocessor-based system like a Raspberry Pi, at its heart. Many of these can be bought online, often with kits or parts and boards for assembling components. Guiding students to evaluate the appropriateness of the different types of boards and associated kits gave a starting point for students to collect key components. The cost often being less than the price of a textbook. For other components we could purchase and ship them to the group members, as each group had a small budget. This way the project could be built by the members, who were scattered around the country and the world. Usually with one person taking the role of integrator. Online file sharing facilities were used along with different Internet of Things resources. Students conducted meetings and communicated through an online team meeting software and could also talk to instructors and teaching assistants live through the meeting tools. Projects developed in somewhat of a similar manner, although all were outside the laboratory and the usual test equipment and electronic board prototyping facilities.

How was the student's learning impacted and what could be optimized through these adjustments? In an attempt to examine and understand the changes a situated learning framework could be considered. The learning environment, the communication between people within the course, the tools used (as well as the lack of tools) could all be considered and comparisons could be made with the 'normal' laboratory environment where student's worked. A colleague had a similar course in another engineering department that focused on computer system and software development. We worked together to consider the different situated learning environments. The results were published in the proceeding of a conference where the work had been presented [251]. In refining the theoretical aspect of the analysis we also used the Kolb learning cycle to consider how the student's learning from their project work could be described through the model. The advantage of this was that it was possible to acknowledge the differences but also to consider how to enhance the learning through emphasis and perhaps adjustment of the facilities.

To approach the examination of the changes of moving from in-person to online the three elements of situated learning proposed by Johri et al [140] were used; *activities and interactions*, *mediation* and *participation*. These have been used elsewhere in this book. These guide us through the various stages of situated learning and allow us to breakdown the analysis. Further, the use of the four stages of the Kolb learning cycle can allow us to see where there may be emphasis in the the learning. By comparing this emphasis in the two approaches, in-person or online, we can attempt to examine the changes in learning that students may be experiencing. For example, if a laboratory moved from an in-person laboratory build to an online video recording or a simulation-based exercise the key elements of the Kolb cycle for the interaction can be examined, see Table 11.5. Here we can see there is a strong concrete experience (CE) with the lab build, then reflective observation (RO) to see if the build is working correctly and then active experimentation (AE) to modify and correct if necessary. Abstract conceptualization (AC) may be a small or larger activity depending on the complexity of problems encountered with a build. Assume initially this is small (for example, is the power turned on), but it may become large and worthy on inclusion in the interpretation. With regards to the online videoed laboratory there is no construction, only the observing and interpreting the video, so the exercise moves quickly to RO and then more emphasis on AC in trying to fit what is observed to knowledge. It would be the same with a provided simulation though there may be more AE with simulation parameter variations.

Table 11.5 also shows a specific example from the previously mentioned electronics project course, where in years before the pandemic students would assemble a microcontroller board from a kit, soldering it together and then testing it. As students were unlikely to have a soldering station at home the project was adjusted to the students evaluating and choosing a prebuilt board. The result of the exercise is the same, having a working microcontroller board, but the experience is different. The final row in Table 11.3 shows the variation, where the students spend most of their time considering what is needed and which board fits their requirements best and so the majority of work sits in the AC part of the cycle.

As explained in Chapter 2, mediation is the use of the tools that the learner uses and that helps them develop their knowledge and understanding. In many

TABLE 11.5
Kolb cycle interpretations of activities and interactions for two exercises that can be done in the lab or in residence (or remotely). The cycle stages of concrete experience (CE), reflective observation (RO), abstract conceptualization (AC) and active experimentation (AE) are indicated for different activities related to the tasks, whether done in the lab or remotely. Based on [251].

Activity	In lab	In residence
Lab exercise	Lab construction	Video lab or simulation
	CE, undertaking the lab RO, is the lab working AE, modifying and correcting	RO, observing video/sim AC, fitting what is observed to theory AE, simulation parameter variation
Project exercise	Microcontroller board assembly	Constructed microcontroller selection
	CE, soldering RO, testing AE, debugging	AC, deciding what is required

of laboratories that is a key area for learning. With the move to online teaching much of the laboratory equipment is removed. For an electronics project a student may have components, a board for assembly of the components and perhaps a multimeter to measure parameters. Table 11.6 shows the how the situation changes with moving from in a laboratory to at a student's residence. In the '*equipment*' row we can see the Kolb learning cycle stages indicated for considering how the learning is progressing. As there is limited equipment, perhaps batteries instead of a power supply, a multimeter and similar basic tools, the experience of the student will be a little different than in a well equipped lab, with technicians who can supply parts or extra equipment. There can then be an increased RO stage with the evaluation of whether the devices are appropriate and can lead to a consideration of the accuracy of a measurement, which leads into more AC. This is not a situation where professional grade equipment can be used removing many limitations on a measurement or test accuracy. However, with limited equipment and perhaps a challenging test to be made this will have to be considered carefully, perhaps in more consultation with a teaching assistant or an instructor. The limitations can then encourage learning. After consideration there may be a need for further improvement needed and that can lead to AE.

Sometimes the situation is perhaps not too different, see the 'professional software tools' row in Table 11.6. My colleague and co-author encourage the use of software tools that were available for download or online [251]. So working remotely can pose little change unless there are limitations with the

TABLE 11.6
Kolb cycle interpretations for mediation with two types of objects used in the lab or in residence (or remotely). The cycle stages of concrete experience (CE), reflective observation (RO), abstract conceptualization (AC) and active experimentation (AE) are indicated for different activities related to the tasks, whether done in the lab or remotely. Based on [251].

Objects	In lab	In residence
Equipment	Lab equipment	Available tools
	CE, using a variety of equipment RO, assessing measurement AC, interpreting measurement	CE, limited (consumer?) equipment RO, assessing measurement AC, validity of measurement AE, further improvising measurement
Professional software tools	Using tools in the lab	Using tools remotely
	CE, using tool RO, interpreting use AC, understanding value of the tool AE, extended use	CE, using tool RO, interpreting use AC, understanding value of the tool AE, extended use

student's computer and then potentially a remote access to another computer could be a possible work around. Compared to other types of engineering a computational lab is often impacted less, when moving online.

For participation and identity there were also some subtle changes, see Table 11.7. With current online communication tools there is the ability to communicate audio-visually, with microphones and cameras, making synchronous communication possible. Virtual rooms can be set up for groups allowing a team to work together. Email and other text-based chat can help with asynchronous communication. In running the project online a professional level tool was selected for communication and this would provide an experience of working with colleagues remotely and off-site. What was interesting to observe was often a reduced use of cameras by students and sometimes a move away from the communication tool selected but to one that students used more frequently in their social lives. This produced an interesting move away from a professional environment and possibly a blurring with their more everyday lives. This was possibly driven by a need to regularly communicate and the professional tool was not used as often as the alternative tool. Though there was open channels of communication possible with the tools there are differences to working face-to-face in a laboratory. Reluctance to use cameras can lead to a lack of non-verbal feedback in communication. Some groups during

TABLE 11.7
Kolb cycle interpretations for two types of participation that can be done in the lab or in residence (or remotely). The cycle stages of concrete experience (CE), reflective observation (RO) and active experimentation (AE) are indicated for different activities related to the tasks, whether done in the lab or remotely. Based on [251].

Interactions	**In lab**	**In residence**
Group work	In laboratory room	Online
	CE, interaction in the lab RO, effective face to face communication AE, leadership role	CE, interaction using online tools RO, effective online communication AE, adapting synchronous and asynchronous communication
TA and instructor interaction	In person meeting	Online meeting
	CE, in lab discussion/showing RO, interpreting advice/information	CE, video meeting/explaining RO, interpreting use

their laboratory time worked with their cameras always on and equipment in view. Other groups just used audio only, perhaps sometimes due to their network connection bandwidth limitations. The role of the teaching assistants and the instructor changed from being available in the room or moving about a lab room, to dropping into virtual group rooms. One effect of the virtual rooms was to isolate the teams more. In a laboratory the sound of success with part of a project can often have students moving over to the successful team, to observe what is working. The virtual rooms stifled this type of secondary or indirect communication. The project course did use lecture time to have updates and presentations from the groups so all could see what was occurring in the various teams. Though this was at a brief and controlled presentation level, not as a working environment. So the cross-flow of ideas between groups was also reduced, compared to when in the laboratory. Table 11.7 shows how participation and identity could be considered to be adjusted for both group work and the interaction between teaching assistants and instructors. With the latter since the interaction is often in the form of questions or instructions it is a little more direct and perhaps less affected than a project with students working together for a few hours each week.

The consideration of the impact on the feeling of identity is perhaps harder to measure, but it can be considerable. Entering a laboratory and working with professional grade equipment can help with the development of the identity of an engineering. Working from a residence with limited components, or perhaps undertaking a remote laboratory or video lab could reduce the development of an engineering identity.

The purpose of describing this required move to an online project experience is to provide an example of how the situated learning model combined with the Kolb learning cycle can be used to look at the impact of changes to a learning activity. In this case the change was a dramatic one. It shows a benefit to using these frameworks and models of learning, to consider how learning is being changed and hopefully being improved.

Many of the models that have been consider in this book focus on the activities (whether laboratory, classroom, project or co-op activities) and ways to understand those specific activities. At this final stage in the book it is worth zooming out a little broader and considering the whole student experience and how this can be a success for a student. Terenzini and Reason have developed a framework to look at student persistence, Figure 11.3, and consider factors external and internal to the college experience [255, 256]. In their work they define student persistence as the student persisting through to their personal goal in college.

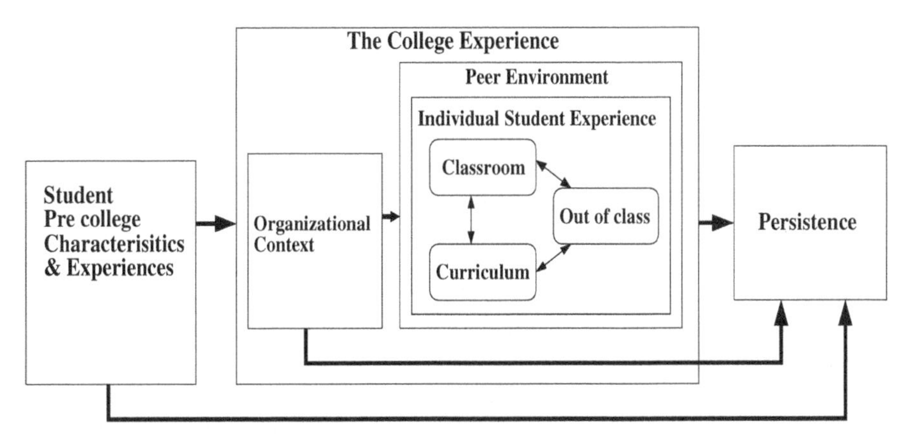

FIGURE 11.3
Terenzini and Reason's model of factors affecting student persistence in college [255, 256].

In this model persistence can be affected by the pre-college experience including socio-demographics, academic preparation and performance and student disposition. Once in college the organizational context can be a factor. Terenzini and Reason highlights 'structural demographic features' and 'organizational behaviour dimensions' [255]. So capturing organizational and systemic features of the college. The student's peer environment is also noted as a factor and can include campus racial climates and the campus academic environments. Finally, the individual student experience is an important factor to the student's persistence and this includes the curriculum experience, the classroom experience (including 'good teaching behaviours' and 'active teaching pedagogies') as well as the out of class experience [219]. The content and details in this book mostly focus on the individual student's experience through the curriculum (use of problem-based learning and CDIO, as well as capstone projects), classroom (e.g. lab experience and in class experience), to out-of-class experience through student societies. Access to co-ops and the support to get a co-op placement or internship can be considered organizational as well as curricular, as could access to support and funding for undergraduate research.

Consideration of Terenzini and Reason's framework is valuable for the instructor and the administrator of an engineering program, as it can provide a clearer picture of how individual elements can affect the overall success of a student. Beneficial experiences can lead to successful outcomes.

11.5 Final Thoughts

The aim through this book has to been to consider the experiential learning in a broad way. Whether it is the more traditionally considered forms of experiential learning, such as co-op work placements or project work, or the less considered learning experiences that can be included into a more traditional classroom. Examples from the literature or my own experience have been given. These examples are just a snapshot of what is out there in the literature. I encourage any educator to look at the current and past literature relating to any course developments they are considering undertaking. There are many resources. It is hoped that this text adds to that body of work and has been helpful in bringing together thoughts and ideas into one resource to help with that search.

As I write this closing the world is slowly emerging from the COVID pandemic which has been the cause of change to not only many people's lives, but also to the way that courses have been delivered in universities and colleges. Because of the nature of the learning, experiential learning has been probably been affected more than traditional lecture-based courses. Students will have worked remotely from their homes on a co-op position. Projects will have moved to remote working and so focused on design with less on construction and testing. Laboratories were video recorded, became remotely accessed or perhaps moved to simulation based. Many things have been learned as faculty overcome this rapid changes that had to be implemented. For myself, I had to adapt courses and I learned more from doing that. The writing of this book at the same time did help me with the adaption and making of changes. Some of the considerations and changes may be for the better and so should be adopted in the courses of the future. The consideration of what worked and what can be improved can be valuable for future learners.

Development of courses can take considerable amount of time and resources. For the future it is hoped that there is more inclusion and adoption of experiential learning. The time it takes for faculty to make these developments needs to be recognized. Tenure and promotion in many universities may still be focused on research productivity. However, there should be a recognition of the scholarly work and planning that can go into the development of a course. For those department heads and administrators that read this it is encouraged that their department and faculty level tenure and promotion guidelines adequately recognize the teaching development work done by a faculty member. Similarly, any experiential learning course needs to be supported and resourced. Champions of existing and new courses need that support to make the developments needed to further the learning of engineering students. Not providing time for faculty or money for funding of equipment can impede the development, and worse damage the enthusiasm of faculty to undertake the development. It is hoped that experiential learning is encouraged and its benefits be recognized.

What then of the future for experiential learning in engineering education? To answer this we need to accept that there may be variations from country to country and accreditation process. In general, there appears to be a growing interest in experiential learning over recent years. This perhaps complements some of the growth in online learning in recent decades, such as with massive open online courses (MOOCs), or with open learning resources. The access to online courses and resources, perhaps associated with micro-credentialing, can work alongside experiential learning to help educate the engineer of the future. For example, online courses could be taken by a student whilst on a co-op work placement.

Looking at specific areas of experiential learning there is likely to be more development away from lecture delivery in courses. Alternative delivery methods could include adoption of peer instruction or flipped classrooms. These approaches are established and known, but perhaps have not yet been widely adopted. Laboratories will continue to be used as a primary way of educating engineers with the practical side of using equipment, construction and analysis of data. Simulation will continue to grow and decisions as to whether have separate courses relating to simulation or to have simulation work replace some of the hardware laboratories will occur. The changes will hopefully be guided by industry and the advisory panels that work with departments of engineering and program coordinators. Along with their development the laboratories still perhaps have a tendency to follow the 'recipe' format that has been long criticized. As new courses and equipment comes along there will be more of a change away from this and perhaps to more investigative and open format laboratories. Perhaps within these laboratories there will be some move to having assessment of competency or level requirements using a clinical examination type form. This could be supporting the assessment of learning outcomes or graduate attributes that accreditation bodies may want to have.

Projects will continue to be in the capstone but may spread into other years. This may even mean that there is a project-based learning that covers each year of an engineering program, which will allow students to develop their learning skills. Capstone projects and even earlier stage project courses may move towards being interdisciplinary. This could be across engineering departments or it could also involve departments of industrial design, architecture, business and the sciences. I recently worked in a multidisciplinary engineering project that looked at bioinspiration within the project topic, which involved mechanical and electrical engineering students. By including a biology professor as a member of the supervisory team there was a true multidisciplinary project with expert input that help with the bioinspiration aspects. When one considers the range of expertise across a university there is perhaps some surprise that there is not more interdisciplinary collaboration in the later years of a program. Leaving the input to programs of different disciplines to core service courses in the early years, such as science and mathematics, seems a missed opportunity.

Undergraduate research may be one area for significant growth in the future. There has been a considerable growth recently, assisted by organizations like CUR. The growth of schemes may be outside of formal degree programs or they may be integrated into and given credit. There may also be the growth of undergraduate research conferences and perhaps journals. Similarly, outside of programs there is going to be the continued growth of student groups and societies. Whether focused on a broad discipline interest, or more specific like unmanned airborne vehicles or AI systems.

It is hoped that diversity, social awareness and ethical responsibility will grow. Indeed with the challenge with the climate crisis there may need to be some more rapid changes in programs to include more education on sustainability and engineering for a sustainable future. Engineers will be at the front of innovation and change to avoid an environmental crisis, and all educators and academic administrators should consider how more can be done. This may lead to the often-encountered dilemma of what to take out if something new is to be included. Perhaps here accreditation can lead and guide. It is interesting to note the change in 2021 of combining the two graduate attributes of 'engineer in society' and 'environment and sustainability' into the new 'the engineer and the world'. This makes an important change and perhaps a strong single graduate attribute from the two previous ones. The use of Fink's taxonomy may be of use to educators with its *caring* and *human dimension* domains. Other social aspects of engineering should not be hidden in this change (and I do not expect they will be). So from an experiential learning perspective there may be more aspects of learning about service learning as well as being aware of social change. For some countries with a colonial background this may need to consider the inclusion of Indigenous peoples and recognition and understanding of their knowledge.

If you read through this book and reached this point I expect that you realize that experiential learning goes beyond running the labs that have run for the past few decades and liaising with the university's co-op office for getting students into relevant work placements. The examples, innovations and publications on experiential learning are clear that this form of learning is a powerful and exciting. It can create meaningful and deep learning as it helps students on their path to become professional engineers.

Bibliography

[1] M. Abdulwahed and Z. K. Nagy. "Towards constructivist laboratory education: Case study for process control laboratory". In: *2008 38th Annual Frontiers in Education Conference.* Oct. 2008, S1B–9–S1B–14.

[2] J.M. Abel. "DOLFFEN: Discovery oriented lab for first-year engineers". In: *Proceedings Frontiers in Education 1995 25th Annual Conference. Engineering Education for the 21st Century.* Vol. 1. Atlanta, GA, USA: IEEE, 1995, pp. 2c4.1–2c4.4. URL: http://ieeexplore.ieee.org/document/483074/.

[3] Alison E. M. Adams, Jocelyn Garcia, and Tinna Traustadóttir. "A quasi experiment to determine the effectiveness of a "Partially Flipped" versus "Fully Flipped" undergraduate class in genetics and evolution". en. In: *CBE—Life Sciences Education* 15.2 (June 2016). Ed. by Erin Dolan, ar11. URL: https://www.lifescied.org/doi/10.1187/cbe.15-07-0157.

[4] Robin S Adams, Jennifer Turns, and Cynthia J Atman. "Educating effective engineering designers: The role of reflective practice". en. In: *Design Studies* 24.3 (May 2003), pp. 275–294. URL: https://linkinghub.elsevier.com/retrieve/pii/S0142694X0200056X.

[5] Guillaume Alinier and Nandini Alinier. "The OSTE: Objective structured technical examination for engineering students". In: *Procs of the Int Conf on Engineering Education.* Gliwice, Poland, July 2005. URL: http://hdl.handle.net/2299/2094.

[6] International Engineering Alliance. *25 years of the Washington Accord: 1989-2014 Celebrating international engineering education standards and recognition.* Tech. rep. Ref 10175. Wellington, New Zealand: International Engineering Alliance, June 2014, p. 24. URL: https://www.ieagreements.org/assets/Uploads/Documents/History/25YearsWashingtonAccord-A5booklet-FINAL.pdf.

[7] International Engineering Alliance. *A History of the International Engineering Alliance and Its Constituent Agreements: Toward Global Engineering Education and Professional Competence Standards.* Tech. rep. Wellington, New Zealand: International Engineering Alliance, Sept. 2015, p. 40. URL: https://www.ieagreements.org/assets/Uploads/Documents/History/IEA-History-1.1-Final.pdf.

[8] International Engineering Alliance. *Graduate Attributes and Professional Competences.* Tech. rep. Version 4. International Engineering Alliance, June 2021, p. 23. URL: https://www.ieagreements.org/assets/Uploads/Documents/IEA-Graduate-Attributes-and-Professional-Competencies-2021.1-Sept-2021.pdf.

[9] International Engineering Alliance. *Washington Accord Signatories.* 2022. URL: https://www.ieagreements.org/accords/washington/signatories/.

[10] A.F. Almarshoud. "The advancement in using remote laboratories in electrical engineering education: A review". en. In: *European Journal of Engineering Education* 36.5 (Oct. 2011), pp. 425–433. URL: http://www.tandfonline.com/doi/abs/10.1080/03043797.2011.604125.

[11] Lorin W. Anderson and David R. Krathwohl, eds. *A taxonomy for learning, teaching, and assessing: A revision of Bloom's taxonomy of educational objectives.* Complete ed. New York: Longman, 2001.

[12] Anonymous. "A mucky business; The Volkswagen scandal". In: *The Economist* 416.8957 (2015), pp. 23–25.

[13] Chris Argyris and Donald A. Schön. *Theory in practice: Increasing professional effectiveness.* 1st ed. San Francisco: Jossey-Bass Publishers, 1974.

[14] ASCE. *ASCE Concrete Canoe Competition.* URL: https://www.asce.org/communities/student-members/conferences/asce-concrete-canoe-competition/.

[15] Carolyn Ash Merkel. "Undergraduate research at the research universities". en. In: *New Directions for Teaching and Learning* 2003.93 (2003), pp. 39–54. URL: https://onlinelibrary.wiley.com/doi/10.1002/tl.87.

[16] BELBIN Associates. *What's Belbin All About?* URL: https://www.belbin.com.

[17] Cynthia J. Atman and Karen M. Bursic. "Verbal protocol analysis as a method to document engineering student design processes". en. In: *Journal of Engineering Education* 87.2 (Apr. 1998), pp. 121–132. URL: https://onlinelibrary.wiley.com/doi/10.1002/j.2168-9830.1998.tb00332.x.

[18] Cynthia J. Atman et al. "Engineering design processes: A comparison of students and expert practitioners". en. In: *Journal of Engineering Education* 96.4 (Oct. 2007), pp. 359–379. URL: https://onlinelibrary.wiley.com/doi/10.1002/j.2168-9830.2007.tb00945.x.

[19] University of Auckland. *MAPTES admissions.* URL: https://www.auckland.ac.nz/en/engineering/study-with-us/maori-and-pacific-at-the-faculty/maptes-admissions.html.

[20] Mercedes Aznar et al. "Interdisciplinary robotics project for first-year engineering degree students". In: *Journal of Technology and Science Education* 5.2 (June 2015), pp. 151–165. URL: http://www.jotse.org/index.php/jotse/article/view/152.

[21] Ken Bain. *What the best college teachers do.* Cambridge, MA: Harvard University Press, 2004.

[22] B. Balamuralithara and P. C. Woods. "Virtual laboratories in engineering education: The simulation lab and remote lab". en. In: *Computer Applications in Engineering Education* 17.1 (Mar. 2009), pp. 108–118. URL: http://doi.wiley.com/10.1002/cae.20186.

[23] Thomas Barrett et al. "A review of university maker spaces". In: *2015 ASEE Annual Conference and Exposition Proceedings.* Seattle, WA: ASEE Conferences, June 2015, pp. 26.101.1–26.101.17. URL: http://peer.asee.org/23442.

[24] Howard S. Barrows. "Problem-based learning in medicine and beyond: A brief overview". en. In: *New Directions for Teaching and Learning* 1996.68 (1996), pp. 3–12. URL: https://onlinelibrary.wiley.com/doi/10.1002/tl.37219966804.

[25] Cheryl Bartlett, Murdena Marshall, and Albert Marshall. "Two-Eyed Seeing and other lessons learned within a co-learning journey of bringing together indigenous and mainstream knowledges and ways of knowing". en. In: *Journal of Environmental Studies and Sciences* 2.4 (Nov. 2012), pp. 331–340. URL: http://link.springer.com/10.1007/s13412-012-0086-8.

[26] Kacey D Beddoes, Brent K Jesiek, and Maura Borrego. "Identifying opportunities for collaborations in international engineering education research on problem- and project-based learning". en. In: *Interdisciplinary Journal of Problem-Based Learning* 4.2 (Sept. 2010). URL: https://docs.lib.purdue.edu/ijpbl/vol4/iss2/3.

[27] F. Bellotti et al. "Serious games and the development of an entrepreneurial mindset in higher education engineering students". en. In: *Entertainment Computing* 5.4 (Dec. 2014), pp. 357–366. URL: https://linkinghub.elsevier.com/retrieve/pii/S1875952114000214.

[28] Marco Bertoni and Alessandro Bertoni. "Measuring experiential learning: An approach based on lessons learned mapping". en. In: *Education Sciences* 10.1 (Dec. 2019), p. 11. URL: https://www.mdpi.com/2227-7102/10/1/11.

[29] Bopaya Bidanda and Richard E. Billo. "On the use of students for developing engineering laboratories". en. In: *Journal of Engineering Education* 84.2 (Apr. 1995), pp. 205–213. URL: http://doi.wiley.com/10.1002/j.2168-9830.1995.tb00167.x.

[30] Peter Bier. “Can students tackle a big problem when time is short?” In: *The CULMS Newsletter* 8 (Dec. 2013). URL: https://www.math.auckland.ac.nz/CULMS/wp-content/uploads/2010/08/CULMS-No-8-December-2013.pdf.

[31] Peter Bier. “Student perceptions of a day-long mathematical modelling group project”. In: *Proceedings of the 2012 AAEE Conference.* Melbourne, Victoria, Australia: Australian Association for Engineering Education, 2012. URL: https://aaee.net.au/wp-content/uploads/2018/10/AAEE2012-Bier.-Day-long_mathematical_modelling_group_project_student_perceptions.pdf.

[32] John B. Biggs. *Teaching for quality learning at university: What the student does.* 2nd ed. OCLC: ocm49976933. Buckingham ; Philadelphia, PA: Society for Research into Higher Education : Open University Press, 2003.

[33] John B. Biggs and Kevin F. Collis. *Evaluating the quality of learning: The SOLO taxonomy (structure of the observed learning outcome).* Educational psychology. New York: Academic Press, 1982.

[34] John B. Biggs and Catherine So-kum Tang. *Teaching for quality learning at university: What the student does.* eng. 4th edition. SRHE and Open University Press imprint. Maidenhead, England New York: McGraw-Hill, Society for Research into Higher Education & Open University Press, 2011.

[35] Gulnara F. Biktagirova and Roza A. Valeeva. “Technological approach to the reflection development of future engineers”. In: *2013 International Conference on Interactive Collaborative Learning (ICL).* Kazan, Russia: IEEE, Sept. 2013, pp. 427–428. URL: http://ieeexplore.ieee.org/document/6644615/.

[36] Benjamin S Bloom, ed. *Taxonomy of educational objectives. 1: Cognitive domain.* eng. 29. print. London: Longman, 1986.

[37] First Nations Financial Management Board. *Closing the Infrastructure Gap.* Tech. rep. 3. West Vancouver, BC: First Nations Financial Management Board, July 2022. URL: https://fnfmb.com/sites/default/files/2022-07/roadmap_project_chapter3_infrastructure_gap_final_v2.pdf.

[38] A. D. Boardman, ed. *Physics programs: A manual of computer exercises for students of physics and engineering.* Chichester [Eng.] ; New York: J. Wiley, 1980.

[39] Cheryl A. Bodnar et al. “Engineers at play: Games as teaching tools for undergraduate engineering students: Research review: Games as teaching tools in engineering”. en. In: *Journal of Engineering Education* 105.1 (Jan. 2016), pp. 147–200. URL: http://doi.wiley.com/10.1002/jee.20106.

[40] David Boehringer and Tilmann Robbe. *LiLa Final Report*. Public report ECP-2008-EDU-428037. Stuttgart: University of Stuttgart, Dec. 2008, p. 21. URL: https://www.lila-project.org/resources/Documents/files/D1-8_annual-public-report_final.pdf.

[41] Marleny M. Bonnycastle and Colin R. Bonnycastle. "Photographs Generate knowledge: Reflections on experiential learning in/outside the social work classroom". en. In: *Journal of Teaching in Social Work* 35.3 (May 2015), pp. 233–250. URL: http://www.tandfonline.com/doi/full/10.1080/08841233.2015.1027031.

[42] Stephen J. Bordes et al. "Towards the optimal use of video recordings to support the flipped classroom in medical school basic sciences education". en. In: *Medical Education Online* 26.1 (Jan. 2021), p. 1841406. URL: https://www.tandfonline.com/doi/full/10.1080/10872981.2020.1841406.

[43] Lori E.A. Bradford et al. "Co-design of water services and infrastructure for Indigenous Canada: A scoping review". en. In: *FACETS* 3.1 (Oct. 2018). Ed. by Nicholas Vlachopoulos, pp. 487–511. URL: http://www.facetsjournal.com/doi/10.1139/facets-2017-0124.

[44] Bill J. Brooks and Milo D. Koretsky. "The influence of group discussion on students' responses and confidence during peer instruction". en. In: *Journal of Chemical Education* 88.11 (Nov. 2011), pp. 1477–1484. URL: https://pubs.acs.org/doi/abs/10.1021/ed101066x.

[45] Andrew Brown and Hugh Morris. "Development and implementation of a final year civil engineering Capstone Project - Successes, lessons learned, and path forward". In: *2020 ASEE Virtual Annual Conference Content Access Proceedings*. Virtual On line: ASEE Conferences, June 2020, p. 34440. URL: http://peer.asee.org/34440.

[46] John Seely Brown, Allan Collins, and Paul Duguid. "Situated cognition and the culture of learning". In: *Educational Researcher* 18.1 (Jan. 1989), p. 32. URL: http://links.jstor.org/sici?sici=0013-189X%28198901%2F02%2918%3A1%3C32%3ASCATCO%3E2.0.CO%3B2-2&origin=crossref.

[47] Katja Brundiers and Arnim Wiek. "Do we teach what we preach? An international comparison of problem- and project-based learning courses in sustainability". en. In: *Sustainability* 5.4 (Apr. 2013), pp. 1725–1746. URL: http://www.mdpi.com/2071-1050/5/4/1725.

[48] United States Census Bureau. *Quick Facts. United States*. URL: https://www.census.gov/quickfacts/fact/table/US/PST045219.

[49] Engineers Canada. *Public Guideline on the code of ethics*. URL: https://engineerscanada.ca/publications/public-guideline-on-the-code-of-ethics#-the-code-of-ethics.

[50] Govt. of Canada. *Indigenous peoples and communities.* URL: https://www.rcaanc-cirnac.gc.ca/eng/1100100013785/1529102490303.

[51] National Sciences and Engineering Research Council of Canada. *Undergraduate Student Research Awards.* English. Jan. 2022. URL: https://www.nserc-crsng.gc.ca/Students-Etudiants/UG-PC/USRA-BRPC_eng.asp.

[52] Deborah Faye Carter et al. "Co-Curricular connections: The role of undergraduate research experiences in promoting engineering students' communication, teamwork, and leadership skills". en. In: *Research in Higher Education* 57.3 (May 2016), pp. 363–393. URL: http://link.springer.com/10.1007/s11162-015-9386-7.

[53] Ricardo Castedo et al. "Flipped classroom-comparative case study in engineering higher education". en. In: *Computer Applications in Engineering Education* 27.1 (Jan. 2019), pp. 206–216. URL: http://doi.wiley.com/10.1002/cae.22069.

[54] K Cator and M Nichols. *Challenge Based Learning: Take action and make a difference.* Tech. rep. Cupertino, CA: Apple Inc. URL: https://www.apple.com/ca/education/docs/Apple-ChallengedBasedLearning.pdf.

[55] Edward F. Cawley. *The CDIO Syllabus: A statement of goals for undergraduate engineering education.* Tech. rep. CDIO, Jan. 2001. URL: www.cdio.org/files/CDIO_Syllabus_Report.pdf.

[56] Edward F. Cawley et al. "The CDIO Syllabus v2.0: An updated statement of goals for engineering education". In: *Proceedings of the 7th International CDIO Conference.* Technical University of Denmark, Copenhagen, June 2011. URL: http://www.cdio.org/files/project/file/cdio_syllabus_v2.pdf.

[57] Peter Cawley. "The introduction of a problem-based option into a conventional engineering degree course". en. In: *Studies in Higher Education* 14.1 (Jan. 1989), pp. 83–95. URL: http://www.tandfonline.com/doi/abs/10.1080/03075078912331377632.

[58] CDIO. *Project-Based Learning in Engineering Education.* URL: http://www.cdio.org/knowledge-library/project-based-learning.

[59] CEIA. *History of Cooperative Education and Internships.* 2022. URL: https://www.ceiainc.org/about/history/.

[60] *Challenge based learning.* 2022. URL: https://www.challengebased-learning.org.

[61] Cecilia Ka Yuk Chan. "Assessment for community service types of experiential learning in the engineering discipline". en. In: *European Journal of Engineering Education* 37.1 (Mar. 2012), pp. 29–38. URL: http://www.tandfonline.com/doi/abs/10.1080/03043797.2011.644763.

[62] Juebei Chen, Anette Kolmos, and Xiangyun Du. "Forms of implementation and challenges of PBL in engineering education: A review of literature". en. In: *European Journal of Engineering Education* 46.1 (Jan. 2021), pp. 90–115. URL: https://www.tandfonline.com/doi/full/10.1080/03043797.2020.1718615.

[63] Wenqian Chen, Umang Shah, and Clemens Brechtelsbauer. "The discovery laboratory – A student-centred experiential learning practical: Part I – Overview". en. In: *Education for Chemical Engineers* 17 (Oct. 2016), pp. 44–53. URL: https://linkinghub.elsevier.com/retrieve/pii/S1749772816300215.

[64] Monit Cheung and Elena Delavega. "Five-Way experiential learning model for social work education". en. In: *Social Work Education* 33.8 (Nov. 2014), pp. 1070–1087. URL: http://www.tandfonline.com/doi/abs/10.1080/02615479.2014.925538.

[65] Bradley A. Cicciarelli. "Use of pre-recorded video demonstrations in laboratory courses". In: *Chemical Engineering Education* 47.2 (2013), pp. 133–136. URL: https://journals.flvc.org/cee/article/view/118334/116276.

[66] Jerry A Colliver. "Effectiveness of PBL curricula". en. In: *Medical Education* 34.11 (Nov. 2000), pp. 959–960. URL: http://doi.wiley.com/10.1046/j.1365-2923.2000.0818a.x.

[67] CDIO Council. *CDIO standards 3.0.* Tech. rep. CDIO, 2020, p. 14. URL: http://cdio.org/files/CDIO%20STANDARDS%203.pdf.

[68] Genisson Silva Coutinho, Nick A. Stites, and Alejandra J. Magana. "Understanding faculty decisions about the integration of laboratories into engineering education". In: IEEE, Oct. 2017, pp. 1–9. URL: http://ieeexplore.ieee.org/document/8190605/.

[69] CPREE. *Campus Field Guides.* URL: http://cpree.uw.edu/campus-fieldguides/.

[70] CPREE. *CPREE — Consortium to Promote Reflection in Engineering Education.* en-US. URL: http://cpree.uw.edu/.

[71] Michael Davis. "Integrating ethics into technical courses: Micro-insertion". en. In: *Science and Engineering Ethics* 12.4 (Dec. 2006), pp. 717–730. URL: http://link.springer.com/10.1007/s11948-006-0066-z.

[72] Robert H. Davis. "Improving the faculty-student experience in chemical engineering". en. In: *AIChE Journal* 66.5 (May 2020). URL: https://onlinelibrary.wiley.com/doi/10.1002/aic.16960.

[73] Peter J Denning and Matti Tedre. *Computational thinking.* English. OCLC: 1114497325. MIT Press, 2019. URL: http://ieeexplore.ieee.org/xpl/bkabstractplus.jsp?bkn=8709327.

[74] John Dewey. *Experience and education.* 60th anniversary ed. West Lafayette, IN: Kappa Delta Pi, 1998.

[75] Christelle Didier. "Engineering Ethics". In: *A companion to the philosophy of technology.* Ed. by Jan Kyrre Berg Olsen Friis, Stig Andur Pedersen, and Vincent F. Hendricks. Blackwell companions to philosophy 43. OCLC: ocn244766425. Chichester, UK ; Malden, MA: Wiley-Blackwell, 2009.

[76] Shoshana R. Dobrow et al. "A review of developmental networks: Incorporating a mutuality perspective". en. In: *Journal of Management* 38.1 (Jan. 2012), pp. 210–242. URL: http://journals.sagepub.com/doi/10.1177/0149206311415858.

[77] Roisin Donnelly and Marian Fitzmaurice. "Collaborative project-based learning and problem-based learning in higher education: A consideration of tutor and student roles and learner-focused strategies". In: *Emerging Issues in the Practice of University Learning and Teaching.* AISHE 1. All Ireland Society for Higher Education (AISHE), 2005.

[78] Emily Dosmar and B. Audrey Nguyen. "Applying the framework of Fink's taxonomy to the design of a holistic culminating assessment of student learning in biomedical engineering". In: *2021 ASEE Virtual Annual Conference Content Access Proceedings.* Virtual Conference: ASEE Conferences, July 2021, p. 36695. URL: http://peer.asee.org/36695.

[79] Xiangyun Du, Erik de Graaff, and Anette Kolmos, eds. *Research on PBL practice in engineering education.* eng. Rotterdam: Sense Publ, 2009.

[80] Tore Dyba, Neil Maiden, and Robert Glass. "The reflective software engineer: Reflective practice". In: *IEEE Software* 31.4 (July 2014), pp. 32–36. URL: http://ieeexplore.ieee.org/document/6834681/.

[81] Chris Eames and Richard K. Coll. "Cooperative education: Integrating classroom and workplace learning". en. In: *Learning Through Practice.* Ed. by Stephen Billett. Dordrecht: Springer Netherlands, 2010, pp. 180–196. URL: http://link.springer.com/10.1007/978-90-481-3939-2_10.

[82] Chris Eames and Richard K. Coll. "Sociocultural views of learning: A useful way of looking at learning in cooperative education." In: *Journal of Cooperative Education and Internships* 40 (2006), pp. 1–12.

[83] Kristina Edström and Anette Kolmos. "PBL and CDIO: Complementary models for engineering education development". en. In: *European Journal of Engineering Education* 39.5 (Sept. 2014), pp. 539–555. URL: http://www.tandfonline.com/doi/abs/10.1080/03043797.2014.895703.

[84] The Canadian Academy of Engineering. *Engineering education in Canadian Universities.* Tech. rep. Ottawa, Ontario, Canada: The Canadian Academy of Engineering, Sept. 1993. URL: https://www.cae-acg.ca/wp-content/uploads/2014/01/1993_Eng%20Education%20Universities.pdf.

[85] American Society for Engineering Education. *Engineering and Engineering Technology by the Numbers, 2019.* Tech. rep. Washington, D.C.: American Society for Engineering Education, 2020. URL: https://ira.asee.org/wp-content/uploads/2021/06/Engineering-by-the-Numbers-2019-JUNE-2021.pdf.

[86] Noel Entwistle, Maureen Hanley, and Dai Hounsell. "Identifying distinctive approaches to studying". en. In: *Higher Education* 8.4 (July 1979), pp. 365–380. URL: http://link.springer.com/10.1007/BF01680525.

[87] Laura Ettinger, Nicole Conroy, and William Barr. "What late-career and retired women engineers tell us: Gender challenges in historical context". en. In: *Engineering Studies* 11.3 (Sept. 2019), pp. 217–242. URL: https://www.tandfonline.com/doi/full/10.1080/19378629.2019.1663201.

[88] Katherine Evans et al. "Ethical decision making in autonomous vehicles: The AV ethics project". en. In: *Science and Engineering Ethics* 26.6 (Dec. 2020), pp. 3285–3312. URL: http://link.springer.com/10.1007/s11948-020-00272-8.

[89] Association of Experiential Education. *What Is Experiential Education?* URL: https://www.aee.org/what-is-ee.

[90] Institute for Experiential Learning. *Kolb Learning Style Inventory 4.0.* 2021. URL: https://experientiallearninginstitute.org/programs/assessments/kolb-learning-style-inventory-4-0/.

[91] Institute for Experiential Learning. *Learning Styles.* 2021. URL: https://experientiallearninginstitute.org/resources/learning-styles/.

[92] Wendy Faulkner. "'Nuts and Bolts and People': Gender-Troubled engineering identities". en. In: *Social Studies of Science* 37.3 (June 2007), pp. 331–356. URL: http://journals.sagepub.com/doi/10.1177/0306312706072175.

[93] Lyle D. Feisel and Albert J. Rosa. "The role of the laboratory in undergraduate engineering education". en. In: *Journal of Engineering Education* 94.1 (Jan. 2005), pp. 121–130. URL: http://doi.wiley.com/10.1002/j.2168-9830.2005.tb00833.x.

[94] Richard M Felder and Linda K Silverman. “Learning and teaching styles in engineering education”. en. In: *Engineering Education* 78.7 (1988), pp. 678–681. URL: https://www.engr.ncsu.edu/wp-content/uploads/drive/1QP6kBI1iQmpQbTXL-08HSl0PwJ5BYnZW/1988-LS-plus-note.pdf.

[95] H. Fernandes et al. “Remote real laboratories in massive open on-line laboratories (MOOLs): A live demonstration at experimenta@2015”. In: IEEE, June 2015, pp. 95–96. URL: http://ieeexplore.ieee.org/document/7463223/.

[96] Susan Fineran et al. “Sharing Common ground: Learning about oppression through an experiential game”. en. In: *Journal of Human Behavior in the Social Environment* 6.4 (Dec. 2002), pp. 1–19. URL: http://www.tandfonline.com/doi/abs/10.1300/J137v06n04_01.

[97] L. Dee Fink. *Creating significant learning experiences: An integrated approach to designing college courses.* Revised and updated edition. Jossey-Bass Higher and Adult Education Series. San Francisco: Jossey-Bass, 2013.

[98] N. D. Finkelstein et al. “When learning about the real world is better done virtually: A study of substituting computer simulations for laboratory equipment”. en. In: *Physical Review Special Topics - Physics Education Research* 1.1 (Oct. 2005). URL: https://link.aps.org/doi/10.1103/PhysRevSTPER.1.010103.

[99] Cindy E. Foor et al. “”We weren’t intentionally excluding them...just old habits”: Women, (lack of) interest and an engineering student competition team”. In: *2013 IEEE Frontiers in Education Conference (FIE)*. Oklahoma City, OK, USA: IEEE, Oct. 2013, pp. 349–355. URL: http://ieeexplore.ieee.org/document/6684846/.

[100] Terri Friel. “Engineering cooperative education: A statistical analysis of employer benefits”. en. In: *Journal of Engineering Education* 84.1 (Jan. 1995), pp. 25–30. URL: https://onlinelibrary.wiley.com/doi/10.1002/j.2168-9830.1995.tb00142.x.

[101] J. E. Froyd, P. C. Wankat, and K. A. Smith. “Five major shifts in 100 years of engineering education”. In: *Proceedings of the IEEE* 100.Special Centennial Issue (May 2012), pp. 1344–1360. URL: http://ieeexplore.ieee.org/document/6185632/.

[102] Cynthia M. Furse and Donna Harp Ziegenfuss. “A Busy Professor’s Guide to Sanely Flipping Your Classroom: Bringing active learning to your teaching practice”. In: *IEEE Antennas and Propagation Magazine* 62.2 (Apr. 2020), pp. 31–42. URL: https://ieeexplore.ieee.org/document/8999522/.

[103] Mohamed Galaleldin et al. "The impact of makerspaces on engineering education". In: *Proceedings of the Canadian Engineering Education Association (CEEA)* (Jan. 2017). URL: https://ojs.library.queensu.ca/index.php/PCEEA/article/view/6481.

[104] Benoît Galand, Mariane Frenay, and Benoît Raucent. "Effectiveness of problem-based learning in engineering education: A comparative study on three levels of knowledge structure". In: *The International Journal of Engineering Education* 28.Extra 4 (2012), pp. 939–947. URL: https://dial.uclouvain.be/pr/boreal/object/boreal:111180/datastream/PDF_01/view.

[105] Adam Gaudry and Danielle Lorenz. "Indigenization as inclusion, reconciliation, and decolonization: Navigating the different visions for indigenizing the Canadian Academy". en. In: *AlterNative: An International Journal of Indigenous Peoples* 14.3 (Sept. 2018), pp. 218–227. URL: http://journals.sagepub.com/doi/10.1177/1177180118785382.

[106] David Gelles. "Boeing 737 Max: What's happened after the 2 deadly crashes". In: *New York Times* (Oct. 2019). URL: https://www.nytimes.com/interactive/2019/business/boeing-737-crashes.html.

[107] Robert Gerlick et al. "Assessment structure and methodology for design processes and products in engineering Capstone courses". In: *2008 Annual Conference & Exposition Proceedings*. Pittsburgh, Pennsylvania: ASEE Conferences, June 2008, pp. 13.240.1–13.240.21. URL: http://peer.asee.org/4110.

[108] James G. Greeno, Allan M. Collins, and Lauren B. Resnick. "Cognition and learning". Ed. by David C Berliner and Robert C Calfee. In: *Handbook of educational psychology*. New York: Simon & Schuster Macmillan, 1996.

[109] Patrick J. Griffis. "Information literacy in business education experiential learning programs". en. In: *Journal of Business & Finance Librarianship* 19.4 (Oct. 2014), pp. 333–341. URL: http://www.tandfonline.com/doi/abs/10.1080/08963568.2014.952987.

[110] Kursty Groves, Edward Denison, and Will Knight. *I wish I worked there! a look inside the creative spaces of the world's most famous brands*. eng. Chichester, West Sussex: Wiley, 2010.

[111] Aida Guerra, Ronald Ulseth, and Anette Kolmos. *PBL in engineering education: International perspectives on curriculum change*. English. OCLC: 1004225489. 2017. URL: https://search.ebscohost.com/login.aspx?direct=true&scope=site&db=nlebk&db=nlabk&AN=1594220.

[112] Marilyn R. Gugliucci and Audrey Weiner. "Learning by living: Life-Altering medical education Through nursing home-based experiential learning". en. In: *Gerontology & Geriatrics Education* 34.1 (Jan. 2013), pp. 60–77. URL: http://www.tandfonline.com/doi/abs/10.1080/02701960.2013.749254.

[113] R M Harden et al. "Assessment of clinical competence using objective structured examination." en. In: *BMJ* 1.5955 (Feb. 1975), pp. 447–451. URL: http://www.bmj.com/cgi/doi/10.1136/bmj.1.5955.447.

[114] Ronald M. Harden and Robert G. Cairncross. "Assessment of practical skills: The objective structured practical examination (OSPE)". en. In: *Studies in Higher Education* 5.2 (Jan. 1980), pp. 187–196. URL: http://www.tandfonline.com/doi/abs/10.1080/03075078012331377216.

[115] Ronald M. Harden, Patricia Lilley, and Madalena Patricio. *The definitive guide to the OSCE: The Objective Structured Clinical Examination as a performance assessment.* Edinburgh ; New York: Elsevier, 2016.

[116] James G. Harris et al. "Journal of engineering education round table: Reflections on the Grinter Report". en. In: *Journal of Engineering Education* 83.1 (Jan. 1994), pp. 69–94. URL: http://doi.wiley.com/10.1002/j.2168-9830.1994.tb00120.x.

[117] M Harvey, D Coulson, and A McMaugh. "Towards a theory of the ecology of reflection: Reflective practice for experiential learning in higher education." In: *Journal of University Teaching & Learning Practice* 13.2 (2016). URL: https://ro.uow.edu.au/jutlp/vol13/iss2/2.

[118] Marina Harvey et al. "Aligning reflection in the cooperative education curriculum." In: *International Journal of Work-Integrated Learning* 11.3 (2010), p. 137.

[119] Annamarie Hatcher et al. "Two-Eyed seeing in the classroom environment: Concepts, approaches, and challenges". en. In: *Canadian Journal of Science, Mathematics and Technology Education* 9.3 (Sept. 2009), pp. 141–153. URL: http://www.tandfonline.com/doi/abs/10.1080/14926150903118342.

[120] David R. Haws. "Ethics instruction in engineering education: A (Mini) meta-analysis". en. In: *Journal of Engineering Education* 90.2 (Apr. 2001), pp. 223–229. URL: https://onlinelibrary.wiley.com/doi/10.1002/j.2168-9830.2001.tb00596.x.

[121] Kent Hecker et al. "Assessment of First-Year Veterinary Students' clinical skills using objective structured clinical examinations". en. In: *Journal of Veterinary Medical Education* 37.4 (Dec. 2010), pp. 395–402. URL: https://jvme.utpjournals.press/doi/10.3138/jvme.37.4.395.

[122] Graham D. Hendry, Miriam Frommer, and Richard A. Walker. "Constructivism and problem-based learning". en. In: *Journal of Further and Higher Education* 23.3 (Oct. 1999), pp. 369–371. URL: http://www.tandfonline.com/doi/abs/10.1080/0309877990230306.

[123] Justin L. Hess and Grant Fore. "A systematic literature review of US engineering ethics interventions". en. In: *Science and Engineering Ethics* (Apr. 2017). URL: http://link.springer.com/10.1007/s11948-017-9910-6.

[124] Mariana G. Hewson et al. "Integrative medicine: Implementation and evaluation of a professional development program using experiential learning and conceptual change teaching approaches". en. In: *Patient Education and Counseling* 62.1 (July 2006), pp. 5–12. URL: https://linkinghub.elsevier.com/retrieve/pii/S0738399106000851.

[125] Rosemary Hill et al. "Working with Indigenous, local and scientific knowledge in assessments of nature and nature's linkages with people". en. In: *Current Opinion in Environmental Sustainability* 43 (Apr. 2020), pp. 8–20. URL: https://linkinghub.elsevier.com/retrieve/pii/S1877343519301447.

[126] Masako Hosaka. "Women's experiences in the engineering laboratory in Japan". en. In: *European Journal of Engineering Education* 39.4 (July 2014), pp. 424–431. URL: http://www.tandfonline.com/doi/abs/10.1080/03043797.2014.883363.

[127] Nathan Hotaling et al. "A quantitative analysis of the effects of a multidisciplinary engineering Capstone design course". en. In: *Journal of Engineering Education* 101.4 (Oct. 2012), pp. 630–656. URL: https://onlinelibrary.wiley.com/doi/10.1002/j.2168-9830.2012.tb01122.x.

[128] Susannah Howe. "Where are we now? Statistics on Capstone courses nationwide". In: *Advances in Engineering Education* 2.1 (2010).

[129] Susannah Howe, Laura Rosenbauer, and Sophia Poulos. "The 2015 Capstone design survey results: Current practices and changes over time". en. In: *International Journal of Engineering Education* 33.5 (2017), pp. 1393–1421.

[130] Cynthia Hsieh and Lorrie Knight. "Problem-based learning for engineering students: An evidence-based comparative study". en. In: *The Journal of Academic Librarianship* 34.1 (Jan. 2008), pp. 25–30. URL: https://linkinghub.elsevier.com/retrieve/pii/S0099133307002133.

[131] Woei Hung, David H. Jonassen, and Liu Rude. "Problem-based learning". In: *Handbook of research on educational communication and technology*. Vol. 3. 1. 2008, pp. 485–506.

[132] Shelby D. Hunt and Debra A. Laverie. "Experiential learning and the Hunt-Vitell Theory of ethics: Teaching marketing ethics by integrating theory and practice". en. In: *Marketing Education Review* 14.3 (Oct. 2004), pp. 1–14. URL: https://www.tandfonline.com/doi/full/10.1080/10528008.2004.11488874.

[133] Anne-Barrie Hunter, Sandra L. Laursen, and Elaine Seymour. "Becoming a scientist: The role of undergraduate research in students' cognitive, personal, and professional development". en. In: *Science Education* 91.1 (Jan. 2007), pp. 36–74. URL: https://onlinelibrary.wiley.com/doi/10.1002/sce.20173.

[134] Frank P. Incropera and Robert W. Fox. "Revising a mechanical engineering curriculum: The implementation process". en. In: *Journal of Engineering Education* 85.3 (July 1996), pp. 233–238. URL: http://doi.wiley.com/10.1002/j.2168-9830.1996.tb00238.x.

[135] International Work Group for Indigenous Affairs. *The Indigenous World.* URL: https://www.iwgia.org/en/indigenous-world.html.

[136] Brandon Ingram et al. "A code of ethics for robotics engineers". In: *2010 5th ACM/IEEE International Conference on Human-Robot Interaction (HRI).* Osaka, Japan: IEEE, Mar. 2010, pp. 103–104. URL: http://ieeexplore.ieee.org/document/5453245/.

[137] SAE International. *SAE International's University Programs.* 2022. URL: https://www.sae.org/attend/student-events.

[138] J. W. George Ivany and Malcolm R. Parlett. "The divergent labortory". en. In: *American Journal of Physics* 36.11 (Nov. 1968), pp. 1072–1080. URL: http://aapt.scitation.org/doi/10.1119/1.1974356.

[139] Mark C. James and Shannon Willoughby. "Listening to student conversations during clicker questions: What you have not heard might surprise you!" en. In: *American Journal of Physics* 79.1 (Jan. 2011), pp. 123–132. URL: http://aapt.scitation.org/doi/10.1119/1.3488097.

[140] Aditya Johri, Barbara Olds, and Kevin O'Connor. "Situative frameworks for engineering learning research". In: *Cambridge handbook of engineering education research.* Cambridge University Press, 2014.

[141] Aditya Johri and Barbara M. Olds. "Situated engineering learning: Bridging engineering education research and the learning sciences". en. In: *Journal of Engineering Education* 100.1 (Jan. 2011), pp. 151–185. URL: https://onlinelibrary.wiley.com/doi/10.1002/j.2168-9830.2011.tb00007.x.

[142] David H. Jonassen and Woei Hung. "All problems are not equal: Implications for problem-based learning". In: *Essential readings in problem-based learning : Exploring and extending the legacy of Howard S. Barrows.* Purdue University Press, Jan. 2015, p. 399.

[143] Shawn S. Jordan et al. "Learning from the experiences of Navajo engineers: Looking toward the development of a culturally responsive engineering curriculum". en. In: *Journal of Engineering Education* 108.3 (July 2019), pp. 355–376. URL: https://onlinelibrary.wiley.com/doi/abs/10.1002/jee.20287.

[144] Lina Kantar. "Incorporation of constructivist assumptions into problem-based instruction: A literature review". en. In: *Nurse Education in Practice* 14.3 (May 2014), pp. 233–241. URL: https://linkinghub.elsevier.com/retrieve/pii/S1471595313001698.

[145] Aliye Karabulut-Ilgu, Nadia Jaramillo Cherrez, and Charles T. Jahren. "A systematic review of research on the flipped learning method in engineering education: Flipped Learning in Engineering Education". en. In: *British Journal of Educational Technology* 49.3 (May 2018), pp. 398–411. URL: http://doi.wiley.com/10.1111/bjet.12548.

[146] Kerry K. Karukstis and Council on Undergraduate Research, eds. *Developing and sustaining a research-supportive curriculum: A compendium of successful practices.* eng. Washington D.C: CUR, 2007.

[147] Barbara Kerr. "The flipped classroom in engineering education: A survey of the research". In: Florence, Italy: IEEE, Sept. 2015, pp. 815–818. URL: http://ieeexplore.ieee.org/document/7318133/.

[148] Kamran Z. Khan et al. "The Objective Structured Clinical Examination (OSCE): AMEE Guide No. 81. Part I: An historical and theoretical perspective". en. In: *Medical Teacher* 35.9 (Sept. 2013), e1437–e1446. URL: http://www.tandfonline.com/doi/full/10.3109/0142159X.2013.818634.

[149] Rob Kitchin. "The ethics of smart cities and urban science". en. In: *Philosophical Transactions of the Royal Society A: Mathematical, Physical and Engineering Sciences* 374.2083 (Dec. 2016), p. 20160115. URL: https://royalsocietypublishing.org/doi/10.1098/rsta.2016.0115.

[150] David A. Kolb. *Experiential learning: Experience as the source of learning and development.* Englewood Cliffs, NJ: Prentice-Hall, 1984.

[151] Anette Kolmos. "Reflections on project work and problem-based learning". en. In: *European Journal of Engineering Education* 21.2 (June 1996), pp. 141–148. URL: https://www.tandfonline.com/doi/full/10.1080/03043799608923397.

[152] Susan R. Komives et al. "A leadership identity development model: Applications from a grounded theory". en. In: *Journal of College Student Development* 47.4 (2006), pp. 401–418. URL: http://muse.jhu.edu/content/crossref/journals/journal_of_college_student_development/v047/47.4komives.html.

[153] Tiina M. Komulainen et al. "Experiences on dynamic simulation software in chemical engineering education". en. In: *Education for Chemical Engineers* 7.4 (Dec. 2012), e153–e162. URL: https://linkinghub.elsevier.com/retrieve/pii/S1749772812000152.

[154] Milo Koretsky, Christine Kelly, and Edith Gummer. "Student perceptions of learning in the laboratory: Comparison of industrially situated virtual laboratories to Capstone physical laboratories". en. In: *Journal of Engineering Education* 100.3 (July 2011), pp. 540–573. URL: http://doi.wiley.com/10.1002/j.2168-9830.2011.tb00026.x.

[155] David R. Krathwohl. "A revision of bloom's taxonomy: An overview". In: *Theory into Practice* 41.4 (Nov. 2002), pp. 212–218. URL: https://doi.org/10.1207/s15430421tip4104_2.

[156] Stephen Laguette. "Assessment of project completion for Capstone design projects". In: *2012 ASEE Annual Conference & Exposition Proceedings*. San Antonio, Texas: ASEE Conferences, June 2012, pp. 25.227.1–25.227.19. URL: http://peer.asee.org/20987.

[157] Shiew Wei Lau, Terence Peng Lian Tan, and Suk Meng Goh. "Teaching engineering ethics using BLOCKS game". en. In: *Science and Engineering Ethics* 19.3 (Sept. 2013), pp. 1357–1373. URL: http://link.springer.com/10.1007/s11948-012-9406-3.

[158] Jean Lave and Etienne Wenger. *Situated learning: Legitimate peripheral participation*. 1st ed. Cambridge University Press, Sept. 1991. URL: https://www.cambridge.org/core/product/identifier/9780511815355/type/book.

[159] L.S. Lee and G. Carter. "A sample survey of Departments of Electrical Engineering to determine recent significant changes in Laboratory Work Pattern at First Year level". In: *International Journal of Electrical Engineering Education* 10 (1973), pp. 131–135.

[160] Chia-Ching Lin and Chin-Chung Tsai. "The relationships between students' conceptions of learning engineering and their preferences for classroom and laboratory learning environments". en. In: *Journal of Engineering Education* 98.2 (2009). _eprint: https://onlinelibrary.wiley.com/doi/pdf/10.1002/j.2168-9830.2009.tb01017.x, pp. 193–204. URL: https://onlinelibrary.wiley.com/doi/abs/10.1002/j.2168-9830.2009.tb01017.x.

[161] E.D. Lindsay and M.C. Good. "Effects of laboratory access modes upon learning outcomes". en. In: *IEEE Transactions on Education* 48.4 (Nov. 2005), pp. 619–631. URL: http://ieeexplore.ieee.org/document/1532371/.

[162] Euan Lindsay and Malcolm Good. "Virtual and distance experiments: Pedagogical alternatives, not logistical alternatives". In: *2006 Annual Conference & Exposition Proceedings.* Chicago, IL: ASEE Conferences, June 2006, pp. 11.1431.1–11.1431.9. URL: http://peer.asee.org/298.

[163] Chung Kwan Lo and Khe Foon Hew. "The impact of flipped classrooms on student achievement in engineering education: A meta-analysis of 10 years of research". en. In: *Journal of Engineering Education* 108.4 (Oct. 2019), pp. 523–546. URL: https://onlinelibrary.wiley.com/doi/abs/10.1002/jee.20293.

[164] Patricia Lucas. "A transactional model to position critical reflection within cooperative education". In: *Asia-Pacific Journal of Cooperative Education* 3 (2017), pp. 257–268. URL: https://openrepository.aut.ac.nz/handle/10292/11068.

[165] Allan Macpherson and Brenton Clark. "Islands of practice: Conflict and a lack of 'Community' in situated learning". en. In: *Management Learning* 40.5 (Nov. 2009), pp. 551–568. URL: http://journals.sagepub.com/doi/10.1177/1350507609340810.

[166] Alejandra J. Magana. "Modeling and simulation in engineering education: A learning progression". en. In: *Journal of Professional Issues in Engineering Education and Practice* 143.4 (Oct. 2017), p. 04017008. URL: http://ascelibrary.org/doi/10.1061/%28ASCE%29EI.1943-5541.0000338.

[167] Salaheddin Malkawi and Omar Al-Araidah. "Students' assessment of interactive distance experimentation in nuclear reactor physics laboratory education". en. In: *European Journal of Engineering Education* 38.5 (Oct. 2013), pp. 512–518. URL: http://www.tandfonline.com/doi/abs/10.1080/03043797.2013.811476.

[168] Yale School of Management. *Top 40 Most Popular Case Studies of 2019.* Feb. 2020. URL: https://som.yale.edu/news/2020/02/top-40-most-popular-case-studies-of-2019.

[169] Michael Manogue and George Brown. "Developing and implementing an OSCE in dentistry". en. In: *European Journal of Dental Education* 2.2 (May 1998), pp. 51–57. URL: http://doi.wiley.com/10.1111/j.1600-0579.1998.tb00039.x.

[170] Sergio Martin et al. "Engineering Societies as a vehicle tool for engineering students". In: *IEEE EDUCON 2010 Conference.* Madrid: IEEE, Apr. 2010, pp. 631–638. URL: http://ieeexplore.ieee.org/document/5492518/.

[171] F. Marton and R. Säljö. "On qualitative differences in learning II outcome as a function of the learner's conception of the task". en. In: *British Journal of Educational Psychology* 46.2 (June 1976), pp. 115–127. URL: http://doi.wiley.com/10.1111/j.2044-8279.1976.tb02304.x.

[172] F. Marton and R. Säljö. "On qualitative differences in learning: Outcome and process*". en. In: *British Journal of Educational Psychology* 46.1 (Feb. 1976), pp. 4–11. URL: http://doi.wiley.com/10.1111/j.2044-8279.1976.tb02980.x.

[173] Veronica Mateos et al. "LiLa booking system: Architecture and conceptual model of a rig booking system for on-line laboratories". In: *International Journal of Online and Biomedical Engineering (iJOE)* 7.4 (Nov. 2011), p. 26. URL: https://online-journals.org/index.php/i-joe/article/view/1837.

[174] Mathworks. *MathWorks Support for Student Competitions.* 2022. URL: https://www.mathworks.com/academia/student-competitions.html.

[175] Mark Matthews and Charles Choi. "Compass course". In: *ASEE Prism* 29.4 (Dec. 2019), pp. 24–29. URL: https://www.asee.org/documents/publications/prism/PrismDec2019.pdf.

[176] Eric Mazur. *Peer instruction: A user's manual.* Prentice Hall series in educational innovation. Upper Saddle River, NJ: Prentice Hall, 1997.

[177] S. McAleer and R. Walker. "Objective structured clinical examination (OSCE)". eng. In: *Occasional Paper (Royal College of General Practitioners)* 46 (Nov. 1990), pp. 39–42.

[178] Bruce McCallum and James C. Wilson. "They said it wouldn't work (A history of cooperative education in Canada)". In: *Journal of Cooperative Education* 24.2–3 (1988), pp. 61–67.

[179] Patricia R. McCarthy and Henry M. McCarthy. "When case studies are not enough: Integrating experiential learning into business curricula". en. In: *Journal of Education for Business* 81.4 (Mar. 2006), pp. 201–204. URL: https://www.tandfonline.com/doi/full/10.3200/JOEB.81.4.201-204.

[180] S.E. McCaslin and M. Young. "Increasing student motivation and knowledge in mechanical engineering by using action cameras and video productions". In: *Advances in Production Engineering & Management* 10.2 (June 2015), pp. 87–96. URL: http://apem-journal.org/Archives/2015/Abstract-APEM10-2_087-096.html.

[181] Roy T. R. McGrann. "Enhancing engineering computer-aided design education using lectures Recorded on the PC". en. In: *Journal of Educational Technology Systems* 34.2 (Dec. 2005), pp. 165–175. URL: http://journals.sagepub.com/doi/10.2190/2B89-MRNQ-WD57-EU48.

[182] B McGurk and T Platton. "Developing objective structured performance related examination (OSPRE)". In: *Policing* 6.2 (1990), pp. 462–468.

[183] Jorge Membrillo-Hernández et al. "Implementation of the challenge-based learning approach in Academic Engineering Programs". en. In: *International Journal on Interactive Design and Manufacturing (IJIDeM)* 15.2-3 (Sept. 2021), pp. 287–298. URL: https://link.springer.com/10.1007/s12008-021-00755-3.

[184] Michael Michie, Michelle Hogue, and Joël Rioux. "The application of both-ways and two-eyed seeing pedagogy: Reflections on engaging and teaching science o post-secondary indigenous students". en. In: *Research in Science Education* 48.6 (Dec. 2018), pp. 1205–1220. URL: http://link.springer.com/10.1007/s11165-018-9775-y.

[185] M. Mina, J. Cowan, and J. Heywood. "Case for reflection in engineering education- and an alternative". In: *2015 IEEE Frontiers in Education Conference (FIE)*. Oct. 2015, pp. 1–6.

[186] J. Mistry and A. Berardi. "Bridging indigenous and scientific knowledge". en. In: *Science* 352.6291 (June 2016), pp. 1274–1275. URL: https://www.sciencemag.org/lookup/doi/10.1126/science.aaf1160.

[187] MIT. *Educational Transformation through Technology at MIT - iLABS.* URL: http://web.mit.edu/edtech/casestudies/ilabs.html.

[188] Tecnológico de Monterrey. *Model Tec21.* URL: https://tec.mx/en/model-tec21.

[189] Eyal Morag et al. "Clinical competence assessment in radiology". en. In: *Academic Radiology* 8.1 (Jan. 2001), pp. 74–81. URL: https://linkinghub.elsevier.com/retrieve/pii/S1076633203807468.

[190] John R. Morelock. "A systematic literature review of engineering identity: Definitions, factors, and interventions affecting development, and means of measurement". en. In: *European Journal of Engineering Education* 42.6 (Nov. 2017), pp. 1240–1262. URL: https://www.tandfonline.com/doi/full/10.1080/03043797.2017.1287664.

[191] Dieter Müller et al. "Mixed reality learning spaces for collaborative experimentation: A challenge for engineering education and training". In: *International Journal of Online and Biomedical Engineering* 3.4 (Nov. 2007). URL: https://online-journals.org/index.php/i-joe/article/view/454.

[192] Paul Nadasdy. "The politics of Tek: Power and the 'Integration' of knowledge." In: *Arctic Anthropology* 32.1/2 (1999), pp. 1–18. URL: www.jstor.org/stable/40316502.

[193] Jeanne L Narum. "A better home for undergraduate science". In: *Issues in Science and Technology* 13.1 (Oct. 1996), pp. 78–84.

[194] United Nations. *United Nations Declaration on the Rights of Indigenous Peoples.* URL: https://www.un.org/development/desa/indigenouspeoples/declaration-on-the-rights-of-indigenous-peoples.html.

[195] Z. Nedic. "Demonstration of collaborative features of remote laboratory NetLab". In: IEEE, July 2012, pp. 1–4. URL: http://ieeexplore.ieee.org/document/6293179/.

[196] Z. Nedic, J. Machotka, and A. Nafalski. "Remote laboratories versus virtual and real laboratories". In: vol. 1. IEEE, 2003, T3E_1–T3E_6. URL: http://ieeexplore.ieee.org/document/1263343/.

[197] Tao Nengfu, Yoong Yuen Soo, and Tan Poh Seng. "CDIO approach for real-time experiental learning of large structures". In: *Proceedings of the 5th International CDIO Conference.* Singapore Polytechnic, Singapore: CDIO, June 2009. URL: http://www.cdio.org/files/document/file/B1.3.pdf.

[198] Flipped Learning Network. *Definition of Flipped Learning.* URL: https://flippedlearning.org/definition-of-flipped-learning/.

[199] Flipped Learning Network. *Flipped Learning Network.* URL: https://flippedlearning.org.

[200] Wendy C. Newstetter and Marilla D. Svinicki. "Learning theories for engineering education practice". In: *Cambridge handbook of engineering education research.* New York: Cambridge University Press, 2014.

[201] Geoffrey R Norman and Henk G Schmidt. "Effectiveness of problem-based learning curricula: Theory, practice and paper darts". en. In: *Medical Education* 34.9 (Sept. 2000), pp. 721–728. URL: http://doi.wiley.com/10.1046/j.1365-2923.2000.00749.x.

[202] Geoffrey R Norman and Henk G Schmidt. "Revisiting 'Effectiveness of problem-based learning curricula: Theory, practice and paper darts'". en. In: *Medical Education* 50.8 (Aug. 2016), pp. 793–797. URL: https://onlinelibrary.wiley.com/doi/10.1111/medu.12800.

[203] Diana Oblinger, ed. *Learning spaces.* eng. An educase e-Book. Washington, DC Boulder, CO: Educause, 2017.

[204] Gilbert O. M. Onwu and Charles Mufundirwa. "A two-eyed seeing context-based approach for incorporating indigenous knowledge into school science teaching". en. In: *African Journal of Research in Mathematics, Science and Technology Education* 24.2 (May 2020), pp. 229–240. URL: https://www.tandfonline.com/doi/full/10.1080/18117295.2020.1816700.

[205] Elisa L. Park and Bo Keum Choi. "Transformation of classroom spaces: Traditional versus active learning classroom in colleges". en. In: *Higher Education* 68.5 (Nov. 2014), pp. 749–771. URL: http://link.springer.com/10.1007/s10734-014-9742-0.

[206] Wiliam Clyde Park. *The cooperative system of education an account of cooperative education as developed in the College of Engineering University of Cincinnati.* Washington: Government Printing Office, 1916. URL: https://eric.ed.gov/?id=ED542633.

[207] John Perkins. *Professor John Perkins' Review of Engineering Skills.* Tech. rep. Department of Business, Innovation and Skills, UK Government, 2013. URL: https://www.raeng.org.uk/publications/other/perkins-review-of-engineering-skills.

[208] J. C. Perrenet, P. A. J. Bouhuijs, and J. G. M. M. Smits. "The suitability of problem-based learning for engineering education: Theory and practice". en. In: *Teaching in Higher Education* 5.3 (July 2000), pp. 345–358. URL: https://www.tandfonline.com/doi/full/10.1080/713699144.

[209] Jacob Perrenet, Ad Aerts, and Jaap van der Woude. "Design based learning in the curriculum of computing science - A skillful struggle." In: Valencia, Spain, July 2003.

[210] Ramana Pidaparti. *Capstone engineering design: Project process and reviews.* English. OCLC: 1264716836. S.l.: MORGAN & CLAYPOOL, 2021.

[211] Raymond Pierotti and Daniel Wildcat. "Traditional ecological knowledge: The third alternative (commentary)". en. In: *Ecological Applications* 10.5 (Oct. 2000), pp. 1333–1340. URL: http://doi.wiley.com/10.1890/1051-0761(2000)010[1333:TEKTTA]2.0.CO;2.

[212] A Piironen et al. "Challenge based learning in engineering education". In: Singapore Polytechnic, Singapore, June 2009, p. 9.

[213] Jennifer Pocock. "Wisdom of the ages". In: *ASEE Prism* Winter (2021).

[214] LiLa Project. *LiLa - Library of Labs.* 2011. URL: https://www.lila-project.org.

[215] Ala Qattawi et al. "A multidisciplinary engineering capstone design course: A case study for design-based approach". en. In: *International Journal of Mechanical Engineering Education* 49.3 (July 2021), pp. 223–241. URL: http://journals.sagepub.com/doi/10.1177/0306419019882622.

[216] J. A. Raelin et al. "Reflection-in-Action on Co-op: The next learning breakthrough". In: *Journal of Cooperative Education and Internship* 42.2 (2008), pp. 9–15. URL: https://wilresearch.uwaterloo.ca/Resource/View/402.

[217] P.K. Raju and Chetan S. Sankar. "Teaching real-world issues through case studies*". en. In: *Journal of Engineering Education* 88.4 (Oct. 1999), pp. 501–508. URL: http://doi.wiley.com/10.1002/j.2168-9830.1999.tb00479.x.

[218] Melissa L Rands and Ann Gansemer-Topf. “"The room itself is active”: How classroom design impacts student engagement”. en. In: *Journal of Learning Spaces* 6.1 (Mar. 2017). URL: http://libjournal.uncg.edu/jls/article/view/1286.

[219] Robert D. Reason. “An examination of persistence research through the lens of a comprehensive conceptual framework”. en. In: *Journal of College Student Development* 50.6 (2009), pp. 659–682. URL: http://muse.jhu.edu/content/crossref/journals/journal_of_college_student_development/v050/50.6.reason.html.

[220] Jay Roberts. “From the Editor: The possibilities and limitations of experiential learning research in higher education”. en. In: *Journal of Experiential Education* 41.1 (Mar. 2018), pp. 3–7. URL: http://journals.sagepub.com/doi/10.1177/1053825917751457.

[221] Carol Rodgers. “Defining reflection: Another look at John Dewey and reflective thinking”. In: *Teachers College Record - TEACH COLL REC* 104 (June 2002), pp. 842–866.

[222] Sarah L. Rodriguez, Charles Lu, and Morgan Bartlett. “Engineering identity development: A review of the higher education literature”. In: *International Journal of Education in Mathematics, Science and Technology* (July 2018), pp. 254–265. URL: http://dergipark.gov.tr/doi/10.18404/ijemst.428182.

[223] Peter Rogers and Richard Freuler. “The "T-Shaped" engineer”. In: *2015 ASEE Annual Conference and Exposition Proceedings*. Seattle, Washington: ASEE Conferences, June 2015, pp. 26.1507.1–26.1507.18. URL: http://peer.asee.org/24844.

[224] Sue V. Rosser. “Group work in science, engineering, and mathematics: Consequences of ignoring gender and race.” In: *College Teaching* 46.3 (1998), pp. 82–88. URL: www.jstor.org/stable/27558891.

[225] Royal Academy of Engineering (Great Britain). *Autonomous systems: Social, legal and ethical issues*. English. OCLC: 495438949. Royal Academy of Engineering, 2009.

[226] Helen E. Rushforth. “Objective structured clinical examination (OSCE): Review of literature and implications for nursing education”. en. In: *Nurse Education Today* 27.5 (July 2007), pp. 481–490. URL: https://linkinghub.elsevier.com/retrieve/pii/S0260691706001389.

[227] Muhammad Syafiq Hazwan Ruslan et al. “Integrated project-based learning (IPBL) implementation for first year chemical engineering student: DIY hydraulic jack project”. en. In: *Education for Chemical Engineers* 35 (Apr. 2021), pp. 54–62. URL: https://linkinghub.elsevier.com/retrieve/pii/S1749772820300658.

[228] Troy D. Sadler. "Situated learning in science education: Socio-scientific issues as contexts for practice". en. In: *Studies in Science Education* 45.1 (Mar. 2009), pp. 1–42. URL: http://www.tandfonline.com/doi/abs/10.1080/03057260802681839.

[229] John Sandars. "The use of reflection in medical education: AMEE Guide No. 44". en. In: *Medical Teacher* 31.8 (Jan. 2009), pp. 685–695. URL: http://www.tandfonline.com/doi/full/10.1080/01421590903050374.

[230] John R. Savery and Thomas M. Duffy. *Problem based learning: An instructional model and its constructivist framework.* Tech. rep. 16-01. Bloomington, IN: University of Indiana, Centre for Research on Learning and Technology, June 2001, p. 17.

[231] Donald A. Schön. *Educating the reflective practitioner: Toward a new design for teaching and learning in the professions.* 1st ed. The Jossey-Bass higher education series. San Francisco: Jossey-Bass, 1987.

[232] Donald A. Schön. *The reflective practitioner: How professionals think in action.* New York: Basic Books, 1983.

[233] Peter Schuster, Andrew Davol, and Joseph Mello. "Student competitions the benefits and challenges". In: *2006 Annual Conference & Exposition Proceedings.* Chicago, IL: ASEE Conferences, June 2006, pp. 11.1155.1–11.1155.11. URL: http://peer.asee.org/1055.

[234] Mieke K. Schuurman, Robert N. Pangborn, and Rick D. McClintic. "Assessing the impact of engineering undergraduate work experience: Factoring in pre-work academic performance". en. In: *Journal of Engineering Education* 97.2 (Apr. 2008), pp. 207–212. URL: https://onlinelibrary.wiley.com/doi/10.1002/j.2168-9830.2008.tb00968.x.

[235] J Seniuk Cicek et al. "Engineering education re-interpreted using the indigenous sacred hoop framework". In: *Proc. of the 8th Research in Engineering Education Symposium: Translating Research into Practice (REES).* Cape Town, South Africa, July 2019, p. 9.

[236] J. Seniuk Cicek et al. "Indigenous initiatives in engineering education in Canada: Collective contributions". In: *Proceedings of the Canadian Engineering Education Association (CEEA)* (June 2020). URL: https://ojs.library.queensu.ca/index.php/PCEEA/article/view/14162.

[237] R.B. Sepe, M. Chamberland, and N. Short. "Web-based virtual engineering laboratory (VE-LAB) for a hybrid electric vehicle starter/alternator". In: vol. 4. IEEE, 1999, pp. 2642–2648. URL: http://ieeexplore.ieee.org/document/799210/.

[238] R.B. Sepe and N. Short. "Web-based virtual engineering laboratory (VE-LAB) for collaborative experimentation on a hybrid electric vehicle starter/alternator". In: *IEEE Transactions on Industry Applications* 36.4 (Aug. 2000), pp. 1143–1150. URL: http://ieeexplore.ieee.org/document/855972/.

[239] Elaine Seymour et al. "Establishing the benefits of research experiences for undergraduates in the sciences: First findings from a three-year study". en. In: *Science Education* 88.4 (July 2004), pp. 493–534. URL: https://onlinelibrary.wiley.com/doi/10.1002/sce.10131.

[240] Shirley Strum Kenny et al. *Reinventing Undergraduate Education: A Blueprint for America's Research Universities.* Tech. rep. State University of New York, Stony Brook, NY, USA: The Boyer Commission on Educating Undergraduates in the Research University, 1998, p. 53. URL: https://eric.ed.gov/?id=ED424840.

[241] Denise R. Simmons, Jennifer Van Mullekom, and Matthew W. Ohland. "The popularity and intensity of engineering undergraduate out-of-class activities". en. In: *Journal of Engineering Education* 107.4 (Oct. 2018), pp. 611–635. URL: https://onlinelibrary.wiley.com/doi/10.1002/jee.20235.

[242] Christine Jane Skilton. "Involving experts by experience in assessing students' readiness to practise: The value of experiential learning in student reflection and preparation for practice". en. In: *Social Work Education* 30.3 (Apr. 2011), pp. 299–311. URL: http://www.tandfonline.com/doi/abs/10.1080/02615479.2010.482982.

[243] Masha Smallhorn. "The flipped classroom: A learning model to increase student engagement not academic achievement". In: *Student Success* 8.2 (July 2017), pp. 43–53. URL: https://studentsuccessjournal.org/article/view/502.

[244] M. K. Smith et al. "Why peer discussion improves student performance on in-class concept questions". en. In: *Science* 323.5910 (Jan. 2009), pp. 122–124. URL: https://www.sciencemag.org/lookup/doi/10.1126/science.1165919.

[245] Gloria Snively and John Corsiglia. "Discovering indigenous science: Implications for science education". en. In: *Science Education* 85.1 (Jan. 2001), pp. 6–34. URL: https://onlinelibrary.wiley.com/doi/10.1002/1098-237X(200101)85:1%3C6::AID-SCE3%3E3.0.CO;2-R.

[246] Caroline J. Speed, Giuseppe A. Lucarelli, and Janet O. Macauley. "Student produced videos - An innovative and creative approach to assessment". In: *International Journal of Higher Education* 7.4 (2018), pp99–109.

[247] S Spierre et al. "An experiential pedagogy for sustainability ethics". In: San Antonio, TX, June 2012. URL: https://peer.asee.org/20921.

[248] Nigel Spinks et al. *Educating engineers for the 21st century: The industry view : A study carried out by Henley Management College for the Royal Academy of Engineering.* English. OCLC: 316743488. Henley-on-Thames: Henley Management College, 2006.

[249] StatsCan. *Aboriginal peoples in Canada: Key results from the 2016 Census.* URL: https://www150.statcan.gc.ca/n1/daily-quotidien/171025/dq171025a-eng.htm?indid=14430-3&indgeo=0.

[250] Alan L Steele. "Vision underwater". In: *Physics Education* 32.6 (Nov. 1997), pp. 387–392. URL: https://iopscience.iop.org/article/10.1088/0031-9120/32/6/011.

[251] Alan L Steele and Cheryl Schramm. "Situated learning perspective for online approaches to laboratory and project work". In: *Proceedings of the Canadian Engineering Education Association (CEEA).* Prince Edward Island: CEEA-ACEG, June 2021. URL: https://ojs.library.queensu.ca/index.php/PCEEA/article/view/14844.

[252] Alan L Steele, Cheryl Schramm, and Kahente Horn-Miller. "Use of an indigenous learning bundle in an engineering project course". In: *Proceedings of the Canadian Engineering Education Association (CEEA)* (June 2020). URL: https://ojs.library.queensu.ca/index.php/PCEEA/article/view/14138.

[253] Camille A. Sutton-Brown. "Photovoice: A methodological guide". en. In: *Photography and Culture* 7.2 (July 2014), pp. 169–185. URL: https://www.tandfonline.com/doi/full/10.2752/175145214X13999922103165.

[254] Tokyo Institute of Technology. *Future & Career.* 2022. URL: https://www.titech.ac.jp/english/admissions/prospective-students/career-paths.

[255] Patrick T. Terenzini and Robert D. Reason. "Parsing the first year of college: A conceptual framework for studying college impacts". In: Philadelphia, PA, Nov. 2005.

[256] Patrick T. Terenzini and Robert D. Reason. "Rethinking between-college effects on student learning: A new model to guide assessment and quality assurance". en. In: *Measuring Quality of Undergraduate Education in Japan.* Ed. by Reiko Yamada. Singapore: Springer Singapore, 2014, pp. 59–73. URL: http://link.springer.com/10.1007/978-981-4585-81-1_4.

[257] Yvonne Tetour, David Boehringer, and Thomas Richter. "Integration of virtual and remote experiments into undergraduate engineering courses". In: IEEE, Oct. 2011, pp. 1–6. URL: http://ieeexplore.ieee.org/document/6086783/.

[258] L. Thigpen et al. "A model for teaching multidisciplinary capstone design in mechanical engineering". In: *34th Annual Frontiers in Education, 2004. FIE 2004.* Savannah, GA, USA: IEEE, 2004, pp. 1204–1209. URL: http://ieeexplore.ieee.org/document/1408741/.

[259] Lian K Ti et al. "Experiential learning improves the learning and retention of endotracheal intubation". en. In: *Medical Education* 43.7 (July 2009), pp. 654–660. URL: http://doi.wiley.com/10.1111/j.1365-2923.2009.03399.x.

[260] Cita van Til and Francy van der Heijden. *PBL study skills an overview.* English. OCLC: 949078084. Maastricht: Department of Educational Development and Research, Universiteit Maastricht, 2006.

[261] Robert H. Todd et al. "A survey of Capstone engineering courses in North America". en. In: *Journal of Engineering Education* 84.2 (Apr. 1995), pp. 165–174. URL: https://onlinelibrary.wiley.com/doi/10.1002/j.2168-9830.1995.tb00163.x.

[262] Karen L. Tonso. "Engineering identity". In: *Cambridge handbook of engineering education research.* New York: Cambridge University Press, 2014.

[263] Karen L. Tonso. "Teams that work: Campus culture, engineer identity, and social interactions". en. In: *Journal of Engineering Education* 95.1 (Jan. 2006), pp. 25–37. URL: http://doi.wiley.com/10.1002/j.2168-9830.2006.tb00875.x.

[264] Joe Tranquillo. "The T-shaped engineer". In: *Journal of Engineering Education Transformations* 30.4 (Apr. 2007), pp. 12–24. URL: https://journaleet.in/articles/the-t-shaped-engineer.

[265] Truth and Reconcilliation Commission of Canada. *Honouring the truth, reconciling for the future: Summary of the final report of the Truth and Reconcilliation Commission of Canada.* English. OCLC: 913832813. 2015. URL: http://epe.lac-bac.gc.ca/100/201/301/weekly_acquisition_lists/2015/w15-24-F-E.html/collections/collection_2015/trc/IR4-7-2015-eng.pdf.

[266] Jennifer Turns, Brook Sattler, and Anette Kolmos. "Designing and refining reflection activities for engineering education". In: *2014 IEEE Frontiers in Education Conference (FIE) Proceedings.* ISSN: 0190-5848, 2377-634X. Oct. 2014, pp. 1–4.

[267] Jennifer Turns et al. "Integrating reflection into engineering education". In: *2014 ASEE Annual Conference & Exposition Proceedings.* Indianapolis, Indiana: ASEE Conferences, June 2014, pp. 24.776.1–24.776.16. URL: http://peer.asee.org/20668.

[268] Council on Undergraduate Research. *Council on Undergraduate Research Issues Updated Definition of Undergraduate Research.* 2021. URL: https://www.cur.org/council_on_undergraduate_research_issues_updated_definition_of_undergraduate_research/.

[269] Council on Undergraduate Research. *Who We Are; Mission and Vision.* 2021. URL: https://www.cur.org/who/organization/mission_and_vision/.

[270] UNESCO and Flavia Schlegel, eds. *UNESCO science report: Towards 2030.* eng. UNESCO science report 2015. OCLC: 942628355. Paris: UNESCO Publ, 2015.

[271] NC State University. *Nuclear reactor program.* 2022. URL: https://nrp.ne.ncsu.edu/internet-reactor-laboratory/.

[272] Faculty of Engineering University of Auckland. *Tuākana Engineering Programme.* URL: https://www.auckland.ac.nz/en/engineering/study-with-us/maori-and-pacific-at-the-faculty/tuakana-engineering-programme.html.

[273] B.P. Van Poppel et al. "Virtual laboratory development for undergraduate engineering courses". In: IEEE, 2004, pp. 644–649. URL: http://ieeexplore.ieee.org/document/1358251/.

[274] Aswin Karthik Ramachandran Venkatapathy. "A study on methodology and implementation of flipped classroom teaching for engineering courses". en. In: *Proceedings of the International Conference on Transformations in Engineering Education.* Ed. by R. Natarajan. New Delhi: Springer India, 2015, pp. 535–540. URL: http://link.springer.com/10.1007/978-81-322-1931-6_61.

[275] Eric Walker, Joseph Pettit, and George Hawkins. *Goals of Engineering Education.* Tech. rep. Jan. 1968. URL: https://www.asee.org/documents/publications/reports/goals_of_engineering_education.pdf.

[276] David Walters and David Zarifa. "Earnings and employment outcomes for male and female postsecondary graduates of coop and non-coop programmes". en. In: *Journal of Vocational Education & Training* 60.4 (Dec. 2008), pp. 377–399. URL: http://www.tandfonline.com/doi/abs/10.1080/13636820802591863.

[277] Graham Walton and Graham Matthews, eds. *Exploring informal learning space in the university: A collaborative approach.* eng. London New York: Routledge, Taylor Francis Group, 2018.

[278] Phillip C. Wankat. "Undergraduate student competitions". en. In: *Journal of Engineering Education* 94.3 (July 2005), pp. 343–347. URL: https://onlinelibrary.wiley.com/doi/10.1002/j.2168-9830.2005.tb00860.x.

[279] Phillip C. Wankat and Frank S. Oreovicz. *Teaching engineering*. New York: McGraw-Hill, 1993.

[280] Theodore F. Wiesner and William Lan. "Comparison of student learning in physical and simulated unit operations experiments". en. In: *Journal of Engineering Education* 93.3 (July 2004), pp. 195–204. URL: https://onlinelibrary.wiley.com/doi/10.1002/j.2168-9830.2004.tb00806.x.

[281] Vincent Wilczynski. "Academic maker spaces and engineering design". In: *2015 ASEE Annual Conference and Exposition Proceedings*. Seattle, WA: ASEE Conferences, June 2015, pp. 26.138.1–26.138.19. URL: http://peer.asee.org/23477.

[282] Jay R Wilson, Thomas T Yates, and Kendra Purton. "Performance, preference, and perception in experiential learning assessment". In: *The Canadian Journal for the Scholarship of Teaching and Learning* 9.2 (Sept. 2018). URL: https://ojs.lib.uwo.ca/index.php/cjsotl_rcacea/article/view/7042.

[283] Jeannette M Wing. "Computational thinking and thinking about computing". en. In: *Philosophical Transactions of the Royal Society A: Mathematical, Physical and Engineering Sciences* 366.1881 (Oct. 2008), pp. 3717–3725. URL: https://royalsocietypublishing.org/doi/10.1098/rsta.2008.0118.

[284] Jeannette M. Wing. "Computational thinking". en. In: *Communications of the ACM* 49.3 (Mar. 2006), pp. 33–35. URL: https://dl.acm.org/doi/10.1145/1118178.1118215.

[285] Kim Graves Wolfinbarger et al. "The influence of engineering competition team participation on students' leadership identity development". en. In: *Journal of Engineering Education* (Aug. 2021), jee.20418. URL: https://onlinelibrary.wiley.com/doi/10.1002/jee.20418.

[286] D. F Wood. "ABC of learning and teaching in medicine: Problem based learning". In: *BMJ* 326.7384 (Feb. 2003), pp. 328–330. URL: https://www.bmj.com/lookup/doi/10.1136/bmj.326.7384.328.

[287] Donald R. Woods. "Problem-based learning for large classes in chemical engineering". en. In: *New Directions for Teaching and Learning* 1996.68 (1996), pp. 91–99. URL: https://onlinelibrary.wiley.com/doi/10.1002/tl.37219966813.

[288] Jingshan Wu, Xiaodong Zou, and Hanbing Kong. "Cultivating T-shaped engineers for 21st century: Experiences in China". In: *2012 ASEE Annual Conference & Exposition Proceedings*. San Antonio, Texas: ASEE Conferences, June 2012, pp. 25.372.1–25.372.10. URL: http://peer.asee.org/21130.

[289] Sarah Yardley, Pim W. Teunissen, and Tim Dornan. "Experiential learning: AMEE Guide No. 63". en. In: *Medical Teacher* 34.2 (Feb. 2012), e102–e115. URL: http://www.tandfonline.com/doi/full/10.3109/0142159X.2012.650741.

[290] Mark R. Young. "Reflection fosters deep learning: The 'Reflection Page & Relevant to You' intervention". In: *Journal of Instructional Pedagogies* 20 (May 2018). URL: https://www.aabri.com/manuscripts/172754.pdf.

[291] Andrew L. Zydney et al. "Faculty perspectives regarding the undergraduate research experience in science and engineering". en. In: *Journal of Engineering Education* 91.3 (July 2002), pp. 291–297. URL: https://onlinelibrary.wiley.com/doi/10.1002/j.2168-9830.2002.tb00706.x.

[292] Andrew L. Zydney et al. "Impact of undergraduate research experience in engineering". en. In: *Journal of Engineering Education* 91.2 (Apr. 2002), pp. 151–157. URL: https://onlinelibrary.wiley.com/doi/10.1002/j.2168-9830.2002.tb00687.x.

Index

R

S

T

U